SING LIKE

SING LIKE FISH

HOW SOUND RULES LIFE UNDER WATER

Amorina Kingdon

CROWN
NEW YORK

Published in the United States by Crown,
an imprint of the Crown Publishing Group,
a division of Penguin Random House LLC, New York.

CROWN and the Crown colophon are registered trademarks of
Penguin Random House LLC.

Library of Congress Cataloging-in-Publication Data
is on file with the publisher.

Hardback ISBN 978-0-593-44277-7
Ebook ISBN 978-0-593-44278-4

Printed in the United States of America on acid-free paper

crownpublishing.com

2 4 6 8 9 7 5 3 1

First Edition

Book design by Belem Bueno

To my family.

TABLE OF CONTENTS

SING LIKE FISH

INTRODUCTION

ONE SUMMER DAY WHEN WE were kids, my brother and I threw our toy trucks off the dock into the lake in front of our house. We watched our miniature yellow graders and cement mixers sink 2 meters to the bottom. Then we jumped in after them. We drove the trucks around the lake bed just as we drove them around the dirt pit and throughout our forested acre in rural Ontario, expanding our domain into the aquatic, between hastily gulped breaths at the surface. We graded pebbles in the shallow fish nests, bulldozed the pondweed and milfoil. We used the digger to load silt clods into the dump truck. At one point, we tried to talk to each other and discovered that, underwater, sound didn't seem to work.

My brother sounded faint and garbled, even when he swam close and screamed a bubble stream. My own voice was loud in my head, but he couldn't hear me. Inevitably, we spent some time shouting all the bad words we knew with impunity.

I remember the rattle of my cement mixer's wheels over the algae-furred rocks was perfectly clear but seemed somehow unconnected to the truck in my hands, as though the sound came from nowhere and all around at once. When a motorboat passed, out in the bay, the outboards that made a lusty buzz in the air were higher-pitched underwater, like a mosquito instead of a hornet. And when Mom's wavering silhouette appeared on the dock, we didn't hear her at all. We could only see her arms moving, beckoning us to (what we could only guess was) lunch.

Underwater trucks was fun, for a day or two. But my brother and I soon hauled our dripping fleet back to the sandpit. We couldn't hear what we needed to, or trust what we did. We couldn't *communicate*, which is essential to a good game of trucks.

This was when I first paid attention to sound underwater. Most people experience something similar. Dunking our heads in the bathtub to rinse shampoo, or frog kicking through a swimming pool, we notice sound is faint, distorted, and apparently contains little useful information. We assume it doesn't *work*. And where we can't sense a world, it's difficult to imagine one exists. For most of human history, our own ears were all we had with which to listen, and human ears are not evolved to work underwater.

IN HIS 1956 film *The Silent World*, Captain Jacques-Yves Cousteau describes in soothing, French-accented English the undersea adventures of the crew of the ship *Calypso*. During World War II, Cousteau had co-developed a regulator, which allowed humans to breathe from pressurized tanks while diving. Pairing scuba (originally an acronym for "Self-Contained Underwater Breathing Apparatus") with underwater videography, Cousteau and his slim, swim-trunk-clad crew, cylinders of air strapped to their backs, swam through coral reefs, among whales, fish, and other creatures, and into the deep. "We have merely skimmed the surface of the ocean," Cousteau narrates to close the film. "Someday we will go much deeper to new discoveries waiting in the silent world." The film became widely known, and its trope of the ocean as a silent world has persisted.

Yet through the nineteenth and twentieth centuries, research driven by warfare, commerce, and curiosity gave rise to technologies like hydrophones, special microphones designed to work in water. Humans began to hear the stunning breadth of aquatic sounds we had been missing. We discovered that for many aquatic animals, while other senses—sight, taste, smell, touch—are often diminished in water, sound becomes enhanced.

As it does above the water, sound carries over distance, in darkness, and around objects. But underwater it travels four and a half times faster than in air, and the right sound under the right conditions can cross seas. Sound holds critical information and mediates vital interactions.

We've learned that whales make more sounds than we'd ever imagined. Some social whales define their groups with unique dialects: Some who have what is arguably "culture"—and debates rage about what this word means for animals and humans—transmit that culture through their calls and even their songs.

We have confirmed that fish can hear, and learned they make many sounds, even daily choruses. Some fish who must find each other to mate drum their swim bladders with some of the fastest muscles in the animal kingdom.

More recently, we've learned how even animals like corals, octopuses, and lobsters, which seem to make few sounds, or have nothing we could call an ear, detect sound beneath the water. Even tiny larvae that must find a hospitable shore, detect the sounds of the coast to find a safe home.

With the help of technology, we've found some animals fraternize in frequencies beyond our perception. Gear designed to sense earthquakes picks up fin whales' low-pitched voices, while dolphins, porpoises, and their toothed-whale brethren peer about the ocean with high-frequency biosonar clicks far above our hearing range, their abilities still unmatched by naval sonar.

We are finding that underwater, sound is the best way to learn about the world, and to communicate, for many animals.

As biologists Hal Whitehead and Luke Rendell write, "The movement of information is the basis of biology. Life happens and creatures evolve because information is transferred." Underwater, information is often—not always, but often—sent and received most accurately, most quickly, over the greatest distances, with sound.

In short: underwater, sound mediates lives.

THIS BOOK SEEKS to explain why sound is so important to animals underwater, how sound behaves differently in water than in air, why we haven't always listened beneath the waves, what we learn when we do, and what we miss when we don't.

Cousteau himself countered the silent-world stereotype in a scene from the 1968 episode of *The Undersea World of Jacques Cousteau* entitled *Savage World of the Coral Jungle*. One scene depicts *Calypso*'s crew music night. While the crew takes up guitars and sings, Cousteau sits apart and listens through headphones to the ocean.

"For me," he narrates, "another kind of music. The sounds from the ocean floor. There are many noises in the silent world: Shrimps, crustaceans, fish, and mammals produce sounds. Except for the sonar signals and loud chatter of sea mammals, the noise level is generally low. Only with a sensitive hydrophone, and [. . .] amplifiers, is it possible to record, identify, and analyze the noises of the sea."

We hear ourselves in the sea too. Humans have become marine mammals of a sort, with our ships, instruments, submarines, and our own sounds underwater. But our voices must seem very strange to animals. Piercing sonar, thudding seismic air guns for geological imaging, bangs from pile drivers, buzzing motorboats, and shipping's broadband growl. We make a lot of noise.

"Noise" is a technical term. It's unwanted sound that interferes with an important acoustic signal. Noise isn't defined by volume or source. The ocean is not and has never been a silent place. But just as animals evolved to live in certain temperatures, or to eat certain food, they also evolved in what may be called certain soundscapes. Human sound underwater is not universally problematic, but it becomes *noise* when it's unwanted.

Globally, shipping noise in the ocean has doubled every decade from 1960 to 2010. The advent of loud seismic surveying and sonar technology has only come in the twentieth century. Many sources became widespread before we understood the soundscapes we were changing.

Much research to date on the impacts of underwater noise has focused on marine mammals, and on acute effects like injury or death. But scientists now study how noise influences the lives of less obviously acoustic animals, such as fish and crabs, scallops, and even seagrass. Because underwater, acoustic space is valuable, and noise is a trespass. We are learning noise impacts communication, mating, fighting, migrating, or bonding in subtle and wide-ranging ways. Sometimes noise is the largest threat to an animal or species, but often it compounds with other threats, such as climate change or pollution.

There are little to no regulations about underwater noise—yet. Internationally, standards are in development, and international organizations are discussing the issue.

Sound underwater is a vast, multifaceted topic. Here, I explore the relationship between sound and animals through a scientific lens, so I neglect discussion of concepts like "song," "language," or "culture" that dip into philosophy or anthropology.

Most important, I acknowledge that coast dwellers and Indigenous communities around the globe have relationships to the sea from time immemorial, and their ways of knowing are deep and profound. Western scientists and Indigenous communities have, in some places, begun long-overdue collaborations using Traditional Ecological Knowledge, and science must acknowledge and respect the unceded land on which their work takes place.

Science is profoundly collaborative. A narrative often centers individuals. In describing pivotal moments of choice or discovery, I shortchange a wide community almost uniformly keen to share credit. These works of research involve teams and their contributions should not be ignored.

Finally, a good deal of the existing science centers on the ocean, and so I offer a salty tale (specifically a temperate and tropical one, though Arctic Ocean ecosystems are a bustling research frontier), though. Yet, fascinating research is done on freshwater lake and river species such as Amazonian piranhas and Malawian cichlids.

Research on sea life and sound is often driven by practical concerns, informing policy decisions like deep-sea mining and Arctic shipping lanes, marine protected areas, and offshore drilling. Yet amid these concerns the emerging data lets us glimpse more and more of acoustic worlds we could never have imagined. What we find is wondrous.

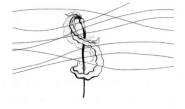

CHAPTER 1

Into a Watery Forest:
Senses in the Sea

I really don't know why it is that all of us are so committed to the
sea, except I think it's because in addition to the fact that the sea
changes, and the light changes, and ships change, it's because we all
came from the sea. And it is an interesting biological fact that all of
us have in our veins the exact same percentage of salt in our blood
that exists in the ocean, and, therefore, we have salt in our blood, in
our sweat, in our tears. We are tied to the ocean. And when we go
back to the sea—whether it is to sail or to watch it—we are going
back from whence we came.

—John F. Kennedy, Remarks at the Dinner for the
America's Cup Crews, September 14, 1962

THE ANCHOR CHAIN RIPS OVER the aluminum bow with a deafening
rattle. When the anchor strikes the bottom of Barkley Sound there's a
sudden silence over the water broken only by the sea's slow wash on the
nearby rocks. The early morning sun hides in September overcast.

"Okay," Kieran Cox says, stretching his arms over his head. "Here we go."

He secures the anchor chain and then bounds toward the boat's stern, over roll-top dry bags and milk crates crammed with neoprene dive gear and surveying equipment. The thirty-three-year-old is ruddy and freckled with a reddish wedge of beard and an athlete's shoulders; and there's a touch of old-school field scientist in his green fisherman's sweater and desert boots, properly laced.

Cox moves aside a meter-tall white PVC pipe stand that he twisted together late last night in his cabin, to which he's lashed two small black cola-can-sized hydrophones. I ask him how he turns them on underwater. "They're on. They're listening right now." He grins and widens his eyes at me.

The *Liber Ero*—Libby for short—is a 6.5-meter aluminum research vessel and dive boat at the Bamfield Marine Sciences Centre, a research campus tucked into Barkley Sound on the west coast of Vancouver Island. Since June, Libby has carried Cox and his colleagues around the sound to two dozen underwater research sites, like this one just off a small rocky islet. More than a hundred such islets dot the sound. Their slopes are forested above the water with British Columbia's characteristic spruce and fir, and beneath the water with kelp.

Kelp are large brown seaweeds, and two species here in the Sound are large enough to form forests, growing up to 30 meters long in towering underwater groves. Bull kelp, or *Nereocystis luetkeana*, is a beautifully simple structure—one long clean bullwhip stalk stretching from a net-like holdfast that grips the rocky bottom to a fist-sized hollow surface float that trails a tuft of long, rubbery blades. Its sleek structure thrives in cool high-energy water wherever waves seethe and crash. In contrast the giant kelp, *Macrocystis pyrifera*, the largest kelp species in the world, sports wrinkled blades all along the stem like a giant cornstalk.

Kelp forests grow along more than a third of the world's coasts, including most of British Columbia's. If you want to understand these temperate coastal ecosystems you need to understand kelp. These forests give structure, shelter, and food to rich groups of plants and animals. But Cox is curious about another service that kelp forests might offer: absorbing unwanted noise and preserving the soundscape.

By his own admission Cox is not an acoustician—a scientist who specializes in the study of sound. (He once described himself to me as merely "sound-curious.") He's an early-career marine ecologist and studies many communities under the waves in addition to kelp, from coral reefs to seagrass beds.

But Cox nonetheless needs to consider sound to understand this kelp community because like light, or temperature, we now know sound is critical for many underwater animals. For this study, his question is: How much unwanted sound—noise—do the great fronds and soft stalks absorb or muffle?

Noise from boats, ships, and other sources is increasing in more and more parts of the ocean, especially near coasts, which in British Columbia often means kelp ecosystems. At the same time, kelp forests themselves are declining. What does that mean for the soundscape in and around these forests? How does noise move through kelp? There are data gaps, and Cox is trying to fill a few.

All summer Cox has been diving into the kelp, where he surveys the forests' inhabitants, sets out the hydrophone stands among the stalks, makes noise nearby, and listens to the recordings. Today in the stern he joins Bridget Maher and Claire Attridge. Cox is now a postdoctorate student at Vancouver's Simon Fraser University, but earned his PhD at the University of Victoria, and still collaborates with his former co-supervisor, Francis Juanes, and lab mates. Maher is the Juanes lab manager, and Attridge is a master's student in the same. Marine stations like Bamfield are often collaboration hotbeds between many researchers, labs, and universities.

They are all cold-water divers, as many marine biologists must be. Their hour-plus-long dives don't allow for wet suits, the standard skintight neoprene, but instead require dry suits—bulky, waterproof garments sealed with stiff sealed rings at the neck and wrists and woolen layers beneath. This makes kitting up on the boat a project.

Cox and Attridge peel off their sweaters and step into the suits with the efficient gestures of long practice. They shrug on the heavy air tanks, attach the requisite hoses. Maher has shaved the nape of her neck so her hair doesn't snag in the tight hood; she French-braids Attridge's hair, hooking the long blond strands with deft fingers. Each day means multiple dives at multiple sites, and the math of scuba safety requires them to take turns so no one spends too much time down. A typical hour-long dive to a depth of 10 meters mandates a break of an hour or so before they can repeat the effort. Cox and Attridge are starting the day off.

Maher records each tank's air levels for her lab records. Cox is bouncing.

"Can I roll off and sit in the water?" he asks rhetorically, sluicing water across Libby's deck as he drops in. Attridge follows. She's carrying orange flagging tape and a clipboard with waterproof paper and a pencil on a string, looking for all the world like a forestry surveyor. In a way, she is. Before each sound experiment, the team surveys the kelp forest for fish and invertebrates, of which there are many in these rich seas.

There's a distant *whoosh* and a pale plume suffuses the air a kilometer away.

"Humpback," Maher says, shading her eyes with her hand.

I've been carrying around a small hydrophone for the past year so I can listen whenever I visit the sea. I drop it overboard and wrestle on my earphones. There's no whale song but I do hear heavy breathing, like someone panting. I realize it's a diver, either Cox or Attridge, though their bubbles riffle the surface dozens of meters away. A testament to sound's underwater range, if you have the gear to listen.

Maher zip-ties more hydrophones to the pipe stands. The divers will carry them down to the kelp forest, placing some at the outer edge fully exposed to the boat noise, the other stands 5 meters back into the fronds with more kelp between them and the sound source. The difference between their respective sound levels will tell Cox how the noise propagates through the forest and how much the kelp is absorbing. Maher caps each stand with a GoPro, to record any fish or other animals visibly reacting to sound.

One noise source Cox uses for this experiment is Libby herself. She's of a size and horsepower with the water taxis, fishing boats, and recreational vessels that coastal British Columbians in these parts use frequently. Cox will drive Libby back and forth past the test site.

The other regional noisemakers are a local ferry, tugs, barges, and a few kilometers out, the shipping lane where cargo and cruise ships pass. Cox wants to know what these behemoths, too, sound like as they pass the kelp forest, but lacking access to such large vessels, he will instead play recordings of their passages from an underwater speaker. It's not perfect, as a speaker can't exactly reproduce the noise of these vessels. But it will provide some data. Maher now hauls this black, frisbee-sized disk from its crate. It's designed to play music underwater for synchronized swimmers and connects to a simple Sony .mp3 player loaded with sound files. Maher tests it, skipping through today's playlist: several pure tones, and an in-situ recording of a boat in these very waters.

Barkley Sound's seafloor hosts one node of an underwater observatory network run by a group at the University of Victoria called Ocean Networks Canada (ONC). The North-East Pacific Time-series Undersea Networked Experiments network (or NEPTUNE; never underestimate scientists' ability to choke a good acronym out of anything) stretches out into the Pacific Ocean from Port Alberni, just north of

Bamfield. The Folger Passage node sits just off Barkley's outermost islands, and consists of two platforms, one in 25 meters of water and one 100 meters down. The boat recording is from a vessel that passed this node nine years ago. When Maher plays the sound to test, it's a mechanical drone that builds slowly with the boat's approach, jet-engineish or vacuum-cleanerish.

Cox and Attridge surface and crawl dripping up Libby's ladder. Attridge's waterproof table is filled out in pencil (impressively neat script, given her thick neoprene gloves). In about thirty-five minutes she's spotted nearly two dozen fish, sea stars, and other creatures. (I parse her shorthand for the names of fish: "Kelp g'ling," "Pile perch," "Red turban.")

Maher hands the divers the hydrophone stands, and they descend to place them. Then, back on the boat, they peel dry suits to the waist, shrug on sweaters, and prepare for the sound test. Cox dunks the synchronized-swimming speaker in the ocean and turns it on. An ominous crackle bursts forth. He wiggles the wires, but clearly water has got past the seals and shorted out the circuits.

"This is why field season should only last so long." Cox runs a hand over his face, allowing himself a brief moment of frustration. Then he makes a decision: They'll just do the noise trials with Libby today, not the speaker. Later he'll repair it with Aquaseal.

They must record one more data point. The noise from a boat or a ship increases with vessel speed, so Cox ties his phone around a wheelhouse post, its camera on video and pointed at the controls, and starts recording the chart plotter, which tracks Libby's pace.

Then the team weighs anchor. Cox turns the wheel with an efficient twist and points her down the narrow channel between our islet and its nearest neighbor. A bump of his hand on the throttle kicks her into gear and we pass the kelp forest, and its hydrophones, at 3 knots (one knot is about 1.8 kilometers per hour). At the channel's end, Cox wheels Libby about and ups her speed. She heaves into a reluctant plane, her wave a foaming V, and we do another pass, faster—and louder. The rising tide has covered several shoals that lie just beneath the surface, so Cox keeps a careful eye on the depth finder.

Cox does another pass, then another, back and forth past the forest, each time testing with more noise. Much of the underwater noise from a boat or ship comes from its propeller, and now Libby's propeller radiates noise into the kelp and past the waiting microphones. How much of this noise is bouncing off the rubbery blades or getting absorbed in the soft stalks of kelp?

After the last pass, Cox grins. "I got up to eighteen knots there," he says.

I met Cox in 2016 on a trip to Calvert Island on BC's Central Coast. He was younger then, a graduate student, but still bullishly driven and peppering conversation with compulsive what-if questions that might one day become experiments. He was first on the dock at 4:30 A.M. to catch the lowest tides for mudflat surveys, uniquely easy to chat with at that ungodly hour.

At some point, squelching through the mud in the predawn gloaming, he mentioned he'd been studying fish sounds. I expressed amazement: I never knew that fish made sound. Cox eagerly informed me that not only do fish make sounds, in some places they *chorus*, like a forest at dawn or dusk. "People say that fish sing like birds," he told me. But strictly speaking, fish evolved millions of years before terrestrial animals, including birds. "Birds," Cox stated, "sing like fish."

THEIR SAMPLING DONE, Maher organizes the gear on the boat, and snacks are shared. I take the opportunity I've been waiting for all day. I squeeze into my own wet suit, wrestle my own mask and snorkel over my head. Then, with two neoprene-gloved fingers I press my mask to my face, lean over *Liber Ero*'s aluminum rail, and fall backward into the Pacific Ocean.

The silver-gray September sky vanishes in a swirl of bubbles as my new sensory world resolves. With my snorkel and wet suit a temporary pass to the underwater world, I experience the ocean with my unaided senses, until quite recently the only way that humans did. Or could.

First I feel: British Columbia's perpetually 10-degree-Celsius seawater trickles at my wet-suit seams, at collarbone, wrist, throat. I wave a hand and touch only cerulean-blue water.

I smell marine funk and taste the salt on my snorkel's rubber mouthpiece.

I hear only my breath.

All I see is haze. Barkley Sound is cloudy with life. There's tiny phytoplankton, the photosynthetic algae that powers the marine food web. There are drifting larvae of dozens of animals and other tiny zooplankton. (Today there's a smattering of molted barnacle skins, translucent eyelash-sized shrimp-like shapes). All of this haze nullifies my sight, and I don't see the giant kelp forest until shadows loom and suddenly I am inside it. I sweep aside wrinkled copper blades, twisting to swim between the bushy stalks. Flashing silver salmon smolts flee before me.

I inhale and kick down. Sight diminishes even more. The light dims quickly to twilight as I touch the gravel bottom 4 meters down. I squint to make out green anemones like fist-sized paint blobs, and spiky purple urchins the size of my head that teeter on their delicate spikes between the kelp stalks.

I strain to listen but I hear nothing.

This is what we think it's like underwater—because this is what our senses tell us. Touch works the same here, friction, pressure, temperature. Orientation is the same too. Smell and taste are the same, carried by chemicals like animal pheromones or human pollution, which lace the water just as they lace the air above. Like most modern humans I lean on sight the most of all my senses, so the obscuring haze and twilight pall is distressing. As for sound, there seems to be none.

But can I trust my senses? When I moved from air to water, I entered a medium that carries information differently. This simple fact has profound implications for how each of my senses works. But more interesting, perhaps, is how much information water itself carries in light, sound, and chemicals for each sense to extract in the first place.

For instance, even if these waters were crystal-clear I would see only a few dozen meters ahead of me. Water absorbs light more quickly than does air, so that even on bright days the sea darkens to twilight and then black not far from the surface. If I could rip off my mask and taste or breathe the sea, I would do so, but water carries chemicals slower than air does.

But sound is different. In fact the water around me shivers with sound that I can't hear. Some is the ocean itself. Bubbles sizzle, currents thrum, all overlain by the white noise of waves. Urchins crunch kelp into the mouths in the center of their underside, softly clicking as they wave their spines to crawl along the bottom. Crabs scrape gravel and clap their claws. Some of the kelp forest dwellers deliberately make sound. Worms snap their jaws. Grunt sculpins, well, grunt when they're scared. Black rockfish, which reside along the seafloor and can live fifty years, make low-pitched hums to impress a mate. The humpback's nearby plume suggests there may be occasional whale calls. Yet I hear none of this.

Historically, human ears have closed us off from underwater sound. Many assumed this perceived silence reflected reality.

My ears are adapted for air; so are my lungs, which begin to burn. I kick for the surface, back to my un-silent world.

WE CARRY THE gear up the gangplank to a shed, hose saltwater off the dry suits, and then before anyone can relax, weigh and measure kelp samples. The team has gathered some stalks and fronds that, quantified and extrapolated, will tell Cox how much actual kelp mass is in the forest. Not all kelp is created equal. Do the narrow bull-kelp stalks absorb less sound than the bushy giant kelp? Do thicker forests absorb more sound than sparser ones? A model will let Cox try to extrapolate this data to other forests on other coasts where kelp forests are changing or dwindling.

Maher, Attridge, and Cox pull the slippery kelp from mesh collection bags and stretch them long on the wooden dock—where they seem dark brown, diminished, and limp. As Cox runs each sample into the shed to weigh it on a small scale, there's good-natured grumbling and gratitude that the field season is almost over, with its long, repetitive days. Then Cox shrugs. "I'm the one who plans these things," he says. "We could have done it with open data. A theory. A model. Would have been boring."

I have trouble imagining someone more allergic to the concept of boring than Cox. He grew up far from the coast in the Okanagan, a warm desert valley in British Columbia's southern interior known for agriculture and wine. By his own admission high school found him a decent athlete but a terrible student.

"I don't think I read one book," he says. He assumed he'd play college basketball and maybe study human kinetics. Then, in 2007, a nineteen-year-old Cox went to Cambodia and tried diving. He loved it. He *really* loved it. He spent the trip working several dive courses from a hut on the beach built over the water, and considered diving professionally. But others cautioned it was hard to make a living as an instructor.

Back in the Okanagan he stuck to his plan and went to college. He joined the local dive rescue team in Okanagan Lake. He saw a lot of fish and collected a lot of golf balls. It was a far cry from Southeast Asia, though he did learn to dive in a dry suit and did a lot of rescue training.

After a semester of college he couldn't stop thinking that he wanted to dive for a living. And the most rewarding way to do that?

"I just leaned fully into the idea that I was going to be a marine biologist," he says. Cox had never applied his considerable drive to academics and now he had to play catch-up.

"It was shitty to be in your twenties, taking Maths 12 [twelfth-grade math]." He laughs.

"But I worked hard. That was the thing. Being a small-town athlete, you develop a *kicker* work ethic. So: at the gym at six forty-five every morning. And I left at six forty-five every evening from school. I learned how to work super hard. Now I grind better than most people."

That work ethic brought Cox to the University of Victoria after two years to finish his undergraduate degree. In his first year there he heard of Prof. Francis Juanes, who had published on underwater sounds. Cox, curious, sent him an email, proposing a project testing how sound affected fish egg development. Juanes liked Cox's ideas and enthusiasm, and the following year, Cox came to Bamfield for the first time. He started a master's degree and then upgraded to a PhD-track program. Now, a decade and a PhD later, he's still diving and still hoping to understand, and so protect, the communities that drew him beneath the waves. And he's still sound-curious. As many marine biologists now do he considers acoustics a fundamental part of an ecosystem.

Kelp measured and gear rinsed, we head uphill through the campus. Bamfield, like many marine stations, is equal parts utilitarian and highbrow. Rusted boat trailers and coils of sea-bleached rope sit in a small parking lot beside a glass-windowed conference room where someone plays a quiet piano. The Canadian flag on the main building is at half-mast for the death of Queen Elizabeth II.

I notice a sculpture on a small knoll. It's a mesh globe about 3 meters high, metal bars running along what would be its longitude lines.

Close up I see the bars are corroded cables as thick as my wrist. Layers of rusted strands twist around one another. It's a nod to the submarine communications cable laid here more than a hundred years ago. The Bamfield Marine Sciences Centre began operating in 1972, when it was called the Bamfield Marine Station, but the buildings date back to 1901–02 when construction began on this eastern landing site of the first telegraph cable to span the Pacific Ocean. The cable was spooled off specialized ships, a colonial project that linked Bamfield to Australia by leapfrogging across British territories in the Pacific. From Bamfield the cable's first leg snaked 6,400 kilometers to Fanning Island (Tabuaeran, in Kiribati). At the time this was the longest cable run in the world.

When the cable was laid ocean depths were measured by painstakingly reeling out long "sounding lines" to the bottom. Each reading took hours. It would be decades before depth finders would supplant these sounding lines—depth finders that use underwater sound. Today, most boats, including Libby, sport one.

THE NEXT MORNING Libby heads southwest through a chilly fog toward Barclay Sound's mouth. A slow mournful honking turns out to be a whistle buoy, its rhythmic notes calling out to mariners the swell's period and marking shoals that await the unwary. We stop in the lee of Bordelais Islet, the outermost rock before the open Pacific. A long belt of bare gray stone girds the islets below the lowest wind-blunted evergreens, marking the astonishing height that Pacific's waves can scour. Of the two local kelp species, it's the high-energy-adapted bull kelp, with its sleek, strong whips, that will dominate here and the surface is speckled with its beribboned floats.

Here also are warm-blooded mammals. Harbor seals lounge on the seaweed-slick rocks. Two sea lions patrol the kelp forest's outer edge, diving up and down like matched coach horses. A single sea otter pokes a curious head above the waves.

Because scientists have studied marine mammals' hearing in more detail than other animals, we know noise can impact them, and so Cox decided not to play the speaker or drive Libby back and forth when these mammals were around, as they are today.

Instead, Attridge and Maher are diving for a temperature logger they placed here back in June. It's a watch-sized gray plastic disk, zip-tied to a white sandbag. The site's waves have almost certainly tumbled the sandbag around the seafloor. They may have to search for it.

Attridge and Maher kit up and topple over Libby's gunwale. Cox directs them periodically.

"Hey, Claire!" he yells, sweeping one arm to the right. "Can you go more that way? That rock that's out of the water, can you do a few passes?" Her distant dark-hooded head nods and descends again. Maher surfaces.

"Claire's headed over to this rock!" Cox yells, gestures.

"If you want to head in that direction as well, I think it's worth checking out." He fidgets, almost visibly vibrating. I have learned, in the past few days, how frequently and frustratingly the sea swallows equipment. I've also learned Cox is intensely proud of his ability to retrieve wayward equipment.

"It's so hard to not just put my dive gear on and jump in the water," he mutters. Instead, he directs his colleagues, who descend for one more try.

It's not immediately obvious what a temperature logger has to do with an underwater sound study.

Back in the 1970s and '80s, scientist Louis Druehl and Parks Canada began mapping and surveying kelp here in Barkley Sound, starting the most long-term and detailed such records on the BC coast. Compared to the older maps, some of today's kelp forests are the same and some have grown, but more have diminished. And a huge reason is temperature. "Definitely as the world's oceans get warmer and we disrupt food webs, kelps can decline," Cox says. They're adapted to certain water temperatures and warming stretches their tolerance.

The average global sea surface temperature is expected to rise several degrees Celsius by the year 2100.

"What happens if it's two to three degrees warmer?" Cox asks as he looks out over the Sound. "There's tons of places where it is already."

Barkley Sound's inner waters are naturally about 2 degrees warmer than outer ocean-facing waters. "It seems to be that kelp farther in the sound is dying back *quite* considerably," Cox says. "And less die-back towards the mouth. Which lines up really nicely with the temperature gradient." Barkley Sound may be a rare glimpse of a warmer future ocean. Its species, its sounds—and its services.

One concept in ecology is that of an "ecosystem service." Mangroves, wetlands, and beaches provide a "service" of protecting coasts from storm surges and waves. Forests clean the air and sequester carbon out of the atmosphere.

Kelp-forest services are many. They absorb wave energy and protect the shore for humans, but they provide services to animals too. They give shelter, food, and habitat for everything from seals to crabs to the mosslike bryozoans that encrust the blades of aquatic plants. In Cox's thinking, muffling boat or ship noise may be another service of BC's vast kelp forests. They preserve the soundscape in which animals are adapted to communicate and listen.

Services are often framed, at least when science gets translated to the public, as a service to people, and expressed in dollar value. It can be a human-centric way to frame nature, sure. But scientists must often do so for research to have an impact.

Policymakers use data like Cox's to make decisions. Perhaps, if he can quantify kelp's noise-absorbing service, and demonstrate how it will change in a warmer ocean, his work can help advise regulators where noise pollution should be limited. Or tell designers of marine protected areas how much boat noise will enter the water if kelp forests are allowed to dwindle. But that requires good noise and temperature data at each of his sites.

That's why Cox really wants to find this logger.

There are almost no enforceable regulations on underwater noise in the world. But that may soon change. The International Maritime Organization (IMO) is updating its ship noise guidelines for member nations. Naval architects are doing cost-effective ship redesigns. Even the International Organization for Standardization (ISO) is creating more guidelines for measuring underwater noise. There's an urgent need for good data.

As a pollutant noise is hard to regulate. It is not visible, like plastic debris. It doesn't linger like spilled oil, and it doesn't sum simply: Two ships do not make twice as much noise as one ship. Not all noises are a problem. The same sound may impact one animal and go unnoticed by another. How to study and measure such a complicated thing, let alone regulate it?

One common tool is a framework of concentric zones of influence, which one can think of like a bull's eye around a noise source. For a very loud source, in the zone very close to the source, the sound could cause physical damage to an animal's ears or organs. An animal in the zone a bit farther away might suffer temporary hearing loss. Farther out still, the noise obscures important sounds, like a mate's grunt or an approaching predator. Past that farther still is the zone of behavior change. The animal reacts to the noise. It flees, moves closer, dives or surfaces, calls louder, or stops calling. The zones are a rough list of ways that noise makes animals do things they wouldn't otherwise do or stop doing or sensing what they need to.

But the zones are limited. In the outer circles impacts become far less obvious and harder to study. Also, most research has been done with marine mammals, but the astute might note that if different species have different hearing abilities, these zones will vary for each source and for each animal that hears it. And different animals perceive sound very differently indeed.

It took scientists a while to untangle the fact that ears like those of marine mammals are only one organ with which underwater animals sense sound. Fish also have ears, but they are profoundly different from mammals'. Invertebrates (and many fish) detect sound not by sensing a sound's pressure changes but the movement of particles—which means most hydrophones measure aspects of sound irrelevant to many animals. (More on this later.)

If anything that perceives sound can be affected by it, I think of those creatures on Attridge's list. Many won't hear Libby as I would. It may not even be correct to call it "hearing."

And our intuition about what an "impact" is, is challenged when we realize how much more important hearing is to many animals underwater than their other senses.

As if this were not complex enough, we get less help than we'd like in understanding sound from acoustic science above the waves. Underwater, sound is *different*. It travels faster and farther, attenuates slower than in air. Depending on its wavelength, it can move differently in shallow water and deep water, and over long distances it can veer and bend, forming silent "shadow zones" or concentrated sound "channels."

So, how did we finally start *really* listening underwater? The scientific story of humans and sound underwater is wide-ranging, from firearm-happy Victoria-era gentleman scientists to whales that sounded uncomfortably like Soviet submarines.

Let me start with a study that first made me question: What *is* hearing?

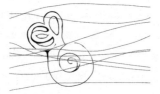

CHAPTER 2

What's in an Ear:
Hearing Underwater

Peeping through my keyhole I see within the range of only about thirty percent of the light that comes from the sun; the rest is infrared and some little ultraviolet, perfectly apparent to many animals, but invisible to me. A nightmare network of ganglia, charged and firing without my knowledge, cuts and splices what I do see, editing it for my brain. Donald E. Carr points out that the sense impressions of one-celled animals are not edited for the brain: "This is philosophically interesting in a rather mournful way, since it means that only the simplest animals perceive the universe as it is."

—Annie Dillard, *Pilgrim at Tinker Creek*

TROPICAL CORAL REEFS TEEM WITH LIFE. Millions of species from the huge to the minuscule are born, live, and die on reefs, including many fish. But some of these fish have a curious life phase, a sort of fishy Rumspringa. These fish breed by releasing their respective gametes to mingle in the water, and then pay no more attention to the resulting tiny larvae, which drift away into the ocean.

This may seem suicidal, but it's actually a safety measure. There's a higher density of predators in the bustling reef community. Baby fish larvae are safer out in the ocean until they grow big enough to hold their own. Then they come back home to the reefs on a dark new moon, ready to transform into adult reef fish. In the southern hemisphere, from Mozambique to the Maldives to Australia, this "recruitment" happens around the November new moon. And until recently scientists had absolutely no idea how far the babies went or how they returned.

Steve Simpson was a grad student in the late 1990s when he set out to study what cued these far-flung larval fish to return. Then, as now, he was slim and athletic with curly brown hair, and he traveled the world studying what, if anything, directed the recruitment of fish larvae back to reefs. The general assumption had been that the fish were so small and the ocean so vast that "they must be at the mercy of ocean currents." They must settle where the sea took them.

Then, in 1999, two studies radically changed his mind. The studies tracked the fishes' journeys using a clever chemical tag and demonstrated that at least half the larvae swam back *to their home reef.* Simpson realized that he was no longer studying a random process. The larvae *chose* their destination, relying on some cue. Some *sense.* So what were the baby fish sensing?

Larval fish do have eyes. And they can taste and smell. They can touch. And they even have little ears. Simpson knew they must be using a sense that acted over distance, so touch and taste were out. Were they reacting to light, sound, or chemicals? A fish can't *see* its home reef (though light seemed to play some role, as recruitment did synchronize to moon phases). Scent tendrils trail through the ocean, but slowly, and are skewed by currents. As for sound, at first Simpson didn't even consider it. Generations had grown up watching Jacques Cousteau's *The Silent World*, which confirmed the human bias that ocean sound wasn't really a thing. That thinking influenced Simpson.

But in the late 1990s, as Simpson was pondering fish senses, a new awareness of sound underwater was getting marine biologists excited.

"There were reports that were starting to be declassified, from the 1950s or so," Simpson says in his enthusiastic, British-accented voice, "the early naval hydrophones, that complained about biological *noise*."

Specifically, World War II and the Cold War had spurred the construction of secret underwater listening networks, and their unprecedented acoustic data led to abundant reports of strange geological and animal sounds. These reports had been classified for decades, but with the end of the Cold War, many non-military marine biologists like Simpson finally got to hear what the ocean sounded like.

THE OCEAN IS never truly silent.

At the lowest frequencies, the ocean rumbles with Earth's seismic mumblings: mud slides in subsea canyons; rock groans as mid-ocean ridges spread; undersea volcanoes roar. In March 2011, a hydrophone in Alaska's Aleutian Islands recorded an unexpected thunder: the Great Honshu earthquake off Japan, 1,500 kilometers away.

Wind and wave sound suffuses the ocean. Wave crests make billions of bubbles that wobble, pop, and collectively form a hissing howl that rises with the wind speed. Raindrops plunging into the sea make more little bubbles. Heavy rain can raise ambient sound levels by 35 decibels. Scientists can track storms, estimate wind speed, or guess raindrop size just by listening to the acoustics. Even falling snow makes a sound described for the first time in 1985, on British Columbia's Cowichan Lake, coincidentally just a hundred kilometers inland from Barkley Sound. One winter day, acoustician Joseph Scrimger was listening to rain with a hydrophone, studying the sounds, when the weather turned to snow. Scrimger kept recording, and when he listened closer, he heard each flake make a soft impact on the water, and then a tiny ultrasonic shriek as the 90 percent of the snowflake that is air released into the water as a miniature bubble.

Ambient wind and rain dominates the open ocean, and hydrophones in these reaches record a haunting, echoing, *big* background sound, while near the coast, currents and crashing waves hiss and sediment skitters back and forth with the surf.

In the Arctic, there's little wave sound beneath the pack ice and landfast sheets. But the sea ice creaks and booms and crashes as it breaks apart. In 1997, scientists with the U.S. National Oceanic and Atmospheric Administration (NOAA) heard a low-frequency sound they famously dubbed the "Bloop." After studying the sounds recorded by hydrophones around Antarctica, they concluded the Bloop was likely an icequake.

In the Southern Ocean icebergs calving from Antartica's glaciers and ice shelves send low-pitched sounds shivering through the ocean and make grinding down-sweeps when they scrape against the seafloor and harmonic tremors when they collide. Scientists acoustically tracked a 55-kilometer-long iceberg dubbed A53a as it drifted from the Weddell Sea in spring and summer 2007, hit a rock 124 meters underwater, spun and ground across the seafloor for six days, got stuck on another shoal, and pinwheeled. Finally it drifted north and broke apart over two months in a quaking cracking symphony.

Coral reef sound travels dozens of kilometers underwater before it fades. Shrimp snap their claws. Parrotfish teeth crunch through coral as they graze on algae. Fish drum their swim bladders and pop their jaws. In the 1970s, Australian scientists recorded reefs doing what they called "chorusing" in the evening as this sound rose to a daily peak.

"A healthy reef," Simpson says, "is a loud reef." Now, with the new declassified reports from wartime hydrophones that proved sound's undersea importance, Simpson wondered: Was sound what the larvae followed home?

It was possible. Hearing tells you things from a distance almost instantly. Sound reflects around objects and travels in the dark, making it a very good sense for warning, orienting, and identifying others. For humans, hearing is the only sense that can trigger a true startle response from a distance, and sound is the most likely distance stimulus to wake you from sleep. Evolutionarily speaking, it's a disadvantage *not* to tap in.

By 1998, Australian scientists had tracked larval fish swimming differently in the presence of reef sounds, but they couldn't link these changes to the *direction* of the sound. The following year, New Zealand researchers attracted baby fish with evening reef chorus recordings. But did the larvae *orient* by the sound? Over what distance? And was the phenomenon widespread? Did all larvae do it?

Simpson went to Lizard Island, partway down the Great Barrier Reef, in November 2003, just before the annual recruitment of the wandering larval fish back to the reefs. He and his colleagues set out light traps—boxes that light up at night—attracting fish like moths to a flame. Using synchronized-swimming speakers, they played "healthy" reef sounds from some boxes and nothing from others, offering the returning larvae a sonic choice. Over three months, 40,161 baby fish returned, and 67 percent went to the noisy traps. A speaker couldn't perfectly mimic natural reef sound, but the baby fish were clearly choosing sound. Simpson and his coworkers' subsequent paper reporting this was charmingly titled "Homeward Sound." Though it still wasn't clear over what exact distance the sound drew the larvae, sound clearly played a big part in recruitment.

But that was far from the end of the story.

THERE ARE OTHER so-called broadcast spawners on the coral reef, with that same open-water phase, such as crabs and even corals themselves. And while fish (and their babies) have ears corals do not. Corals, whose bodies make up the physical structure of reefs, are tiny animals, but "they haven't got a brain, haven't got a central nervous system," Simpson says.

Coral larvae can swim, using tiny hair-like protrusions on their bodies called cilia. They can't make much headway against a current but can beat their cilia just enough to sink or rise. We don't know exactly how far baby corals travel from the reef, compared to baby fish. But somehow they find a good spot and settle down to adulthood.

By 2000, Simpson was collaborating with Netherlands-based researcher Mark Vermeij, whose group studied corals in Curaçao. One student at the lab, whom Simpson co-supervised, was studying how a fish called the French grunt oriented to sound, which meant the group had a common research tool called a choice chamber, which is what it sounds like. It's a circular pool around the rim of which researchers can put stimuli like chemicals or lights. Larvae placed in the center swim to what attracts them, like a puppy in the center of a room moving toward its favorite person. The team had been using the chamber for fish research but, on a whim, the students put *coral* larvae in the chamber and played reef sound.

And the earless larvae moved toward the sound. But how?

Meanwhile, New Zealand researchers were discovering that the larvae of crabs, another animal without ears, moved toward reef sound, and sound even cued them to metamorphose into adults. Again, the distances involved were unclear. But sound clearly played some role.

"So, these invertebrates across a whole range of different taxa, either use sound to avoid landing on a reef and getting eaten, or use sound as a way of getting onto a reef when they're ready to go and make it their home," Simpson recalls. "It just continued to surprise us."

When I first encountered Simpson's work, I was amazed. It seemed a lot of animals underwater were detecting sound without ears. But how? Is that even hearing? And what was that secret wartime listening network all about? To understand why these discoveries about larvae and reef sound made (pardon the pun) such a splash, we must look back—really far back.

IT'S HARD TO understand just how big a billion is. Sure, a billion is a 1 followed by nine 0's, but try to visualize a billion things: A billion pages of this book stacked on top of each other would be about 13 kilometers high. When you're a billion seconds old you're 32.5 years old. The Earth is about 4.5 billion years old. The first simple cells on Earth are thought to have evolved about 3.5 billion years ago, likely from a "primordial soup" of organic molecules that accumulated in the oceans of the young Earth. Over time, the first cells changed and developed new abilities, such as photosynthesis. These new abilities let cells spread and diversify, and eventually gave rise to all the life that we see today. These earliest simple cells wobbled through the sea, consuming and reproducing. These cells' first "senses" were likely touch, and the ability to detect different chemicals in the water. It would take approximately 2 billion years before cells clustered into multi-celled organisms, or metazoans (from the Greek *meta* = "multiple," and *zoa* = "life"), making different sense organs possible.

We don't know much about these first soft, small animals. They didn't have hard parts, like bones or shells, which fossilize and thus leave traces. All we know is that they were in the water, and they moved with structures such as cilia, which are specialized hairlike structures. But then, about 540 million years ago, animal evolution seemed to supercharge, evolving new forms in a relatively brief time in geologic history called the Cambrian Explosion. Large hard-bodied forms of animals with most of the major body plans known today appeared in the fossil record.

Today most scientists group life into six kingdoms: plants, animals, fungi, bacteria, archaea, and protists (these last three are all single-celled). Each kingdom consists of several phyla. A phylum includes different organisms that share important characteristics. Modern animals are divided into around three dozen phyla, many of which first showed up before, during, or not long after the Cambrian Explosion. Porifera (sponges), annelids (segmented worms), mollusks (octopuses, squid, clams), and arthropods (insects, spiders, and crabs), are examples of phyla. Humans are phylum Chordata, along with fish, birds, reptiles, amphibians, and anything else with a nerve cord down its back. Those with a backbone form a group collectively called vertebrates. (The only chordates that are not vertebrates are the tunicates and lancelets, marine animals known as protochordates.)

Reconstructing how these groups are related, not to mention how one body part like an ear evolved, is difficult if you're just using fossils. However, scientists have another tool. By determining the similarities and the differences between the genes of two modern animals, they can tell how recently they had a common ancestor, and thus pinpoint when their respective lineages might have diverged.

The fossil record suggests that by the end of the Cambrian Explosion, our Chordate ancestors had a backbone, and over the next millions of years became fish, and then branched out onto land as amphibians, reptiles, birds, and mammals.

But Chordates are just one animal phylum, which means that most animals on the planet are non-Chordates, what scientists collectively call invertebrates (meaning "without a backbone"). Some 75 percent of all plant and animal species are insects, a suspiciously huge number of which are beetles. Approximately 90 percent of marine organisms are invertebrates.

The bodies of most animals have some sort of symmetry. Chordates, like humans, have bilateral symmetry in which an imaginary plane drawn through the center of the body divides it into two halves that are mirror images of each other. Most animals with bilateral symmetry have a front and rear end with a concentration of nerves and paired sense organs, including ears, at the front end. It's easy to see why Simpson assumed that sound wasn't terribly important to invertebrates, such as crabs and coral, since many invertebrates do not have ears. Post–Cambrian Explosion, modern ears hadn't yet evolved. However, most phyla had specialized structures known as mechanosensory hair cells. Mechanosensory hair cells are similar in appearance to cilia.

It's not clear when these hair cells evolved in invertebrates, or how many times. There are several forms of mechanosensory hair cells, but each variation has a hair-like protrusion (or protrusions) sticking up from the cell. When this "hair," or cilium, is physically bent, the motion sends an electrical impulse down an attached neuron.

In today's marine invertebrates, these hair cells are found in a variety of phyla such as cnidarians, arthropods, and mollusks. When the hair cells are found on the surface of the organism, the motion of water particles bends the cilia, registering stimuli such as the movement of something passing close by.

Some invertebrates' hair cells are also found in organs that sense orientation, balance, and gravity. These organs are called statocysts.

A statocyst consists of a sac-like structure containing a stone or other dense mass (statolith) and numerous innervated sensory hair cells. The statolith's inertia causes it to push against these sensory hairs when the animal moves to and fro. This deflection of the "hair" by the statolith in response to gravity activates neurons, giving feedback to the animal on how it's oriented. And these structures almost certainly allow these animals to sense some sounds.

MOST OF THE information about our surroundings gets to us in the form of waves. Sound is energy that arrives in the form of sound waves. Sound is a pressure wave that can move through anything—air, water, or even solids—anything that can compress even a little bit. A sound wave starts when a source vibrates in a medium such as air, water, or solids. The source bumps some molecules, which bump the molecules next to them, and so on and so on, like those Newton's cradles of hanging balls on C-suite desks. Each molecule moves back into place and gets bumped again by the next vibration as the wave spreads out from the source. (If you want to get technical about it, Merriam-Webster's dictionary defines any wave, sound included, as a "disturbance or variation that transfers energy progressively from point to point.")

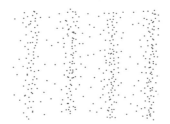

A sound forms waves of compression in the particles of
a medium as it moves, in this case from left to right.

There are several types of wave. Sound is a *longitudinal* wave. This means the particles move back and forth along the wave's direction, like compressions down a slinky. *Transverse* waves happen when particles move at right angles to the wave direction, up and down, like water waves or light. Transverse and longitudinal waves are not the only types of waves found in nature. For example, when an earthquake occurs, particles move in S-shapes or ellipses. But sound is (usually) a longitudinal pressure wave.

A sound's *amplitude* measures how far the particles move back and forth. The farther they move, the higher the amplitude. Generally, human perception of loudness correlates to amplitude.

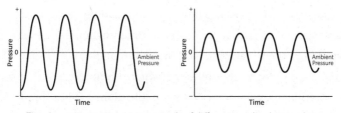

The change in pressure as two sounds of different amplitudes pass by.

Frequency is a measure of how often a vibration occurs. The frequency of something that vibrates and the wave it produces are equal. Frequency is measured in a unit called hertz (Hz), where one hertz is one cycle—one back and forth motion—per second. The greater the speed of the vibrations, the higher the frequency. Sound waves with frequencies below 20 Hz are called infrasound and are not audible to humans. Sound waves with frequencies above 20,000 hertz (kilohertz—kHz) are called ultrasound and are also not audible to humans. (A young human with good ears hears from 20 Hz to about 20 kHz in air. Middle C is 250 Hz; human voices range from around 200 Hz up to several thousand.)

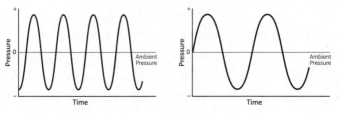

The change in pressure as two sounds of different frequencies pass by.

Hard parts aside, most marine animals' bodies have a density very close to water. So, to a sound, their bodies are almost transparent to sound. Underwater, animals' bodies move with the water.

A strong nearby sound moving through an invertebrate such as a coral polyp or crab may not physically move a dense inert statolith, but since the rest of the animal is moving with the sound, the hair cells' cilia move around the stone, bend, and fire neurons.

In 1950, Cambridge biologist Richard Pumphrey wrote a simple and beautiful paper specifically about the sense of hearing, and addressed whether invertebrates could hear. Gravity and sound both can be turned into nerve impulses and sensed, by physically bending the ciliary bundles of these hair cells. Pumphrey concluded: "It is impossible to read any modern book of comparative physiology without gaining the impression that while most aquatic invertebrates respond to gravity, most of them are deaf. I hope I have made it clear that *the more reasonable conclusion is that an aquatic animal could not respond to gravity unless it heard.*"

Mollusks have these mechanosensory hair cells, or similar ones, on their bodies and in statocysts. Some species also have an abdominal sense organ (ASO) containing sensory hair cells. Some crab species create do-it-yourself statocysts by putting sand into hair-cell-filled pockets in their heads, a handy solution to replacing the mass every time they molt. Crustaceans even detect sound using structures in their joints that also sense proprioception, or their own position movements.

Indeed, nearly all invertebrates on Claire Attridge's census sheet from the Bamfield dive use these tools. Red urchin and sea stars and California sea cucumber. Abalone and giant scallop with their ASOs; umbrella, hermit, and decapod crab with their statocysts and resonant joints. I wonder what I "sounded" like to them, my surface breaths, my flipper strokes, my rustling neoprene.

Modern invertebrates often respond best to nearby vibrations and the particle motion of water. It's possible that sound sensitivity in these animals began to evolve because it was advantageous to sense "touch at a distance," like the slosh of wake from a passing predator or the sound of a coast or a reef, and sensitivity slowly grew from there. It also means, that to understand invertebrate "hearing" we need to consider the difference between sound pressure and particle motion.

Sound waves travel through water by compressing and expanding the water molecules. The amount of compression and expansion determines the sound pressure. The higher the sound pressure, the louder the sound.

Particle motion in water, also known as particle velocity, refers to the actual displacement of water molecules as sound waves travel through the medium. As sound travels it causes water molecules to move back and forth along the direction the wave is traveling. Sound pressure is the primary component that is detected by underwater microphones. Many marine animals, including fish and invertebrates, rely on particle motion to detect and localize sound sources. Measuring particle motion, however, is crucial to understanding how aquatic organisms perceive sound.

This is also a clue as to how baby corals might detect sound underwater, even without an ear. What moves can also be moved. They have cilia on their bodies. These cells are not the same as the primary hair cells of invertebrate balance organs, but the physical principle is similar. The system is known to be sensitive to the particle motions of sound. These cilia could have responded to the reef sounds, and helped guide the coral home.

But what about the baby fish?

The ancestors of fish were primitive vertebrates, and they had hair cells in several body parts. These hair cells had a very different structure from those in invertebrates, and they evolved independently and likely only once. They have an attached neuron. When the "hairs" bend, they generate a nerve impulse, and these along an axis, giving the hair cell's movement *directionality*.

Fish also evolved a special structure called a lateral line that runs the length of their body and is still visible on most species today. Biologists hypothesize that the lateral line began as a row of hair cells that moved in response to touch, and then perhaps also to turbulence in the water, alerting fish to friend and foe.

Meanwhile, inside ancestral fishes' heads, new balance structures evolved. Instead of stone-filled chambers, now there were fluid-filled canals, curled into loops. Fish ancestors eventually evolved three such loops, with each laying in one plane of three-dimensional space. Constrained in the canals, the fluid acts like a sensitive mass. When the fish moved or turned, back or forth or side to side, fluid moved in the canals, and bent the cilia of hair cells in chambers near the base of each loop, sending information from the canals to the brain.

So, the fish's lateral line could sense vibration and perhaps some low-frequency sounds. The loops helped the fish sense acceleration and gravity. But along with the loops, the fish's inner ear also evolved three otolith organs, each of which consisted of a little stone called an otolith contained in chambers lined with patches of hair cells. Like the mass in a statocyst, these otoliths were denser than the fish. When a sound passed by and through the fish, these hair cells moved against the inert otolith, transducing the vibration of sound into a nerve impulse. These early otolith organs probably conveyed relatively low frequency sounds, and were the first structures that evolved primarily dedicated to hearing—the first ear.

SEATTLE'S BURKE MUSEUM of Natural History and Culture houses perhaps the world's largest organized collection of fish otoliths. On a rainy late-November day, Luke Tornabene, the Burke's young and energetic curator of fishes, meets me in a tweed jacket, along with Katherine Maslenikov, the museum's Ichthyology collections manager, tall in a pink sweater. They show me into a lab loud with ventilation pumps and over to a table strewn with cardboard boxes and tiny vials, all holding twin pairs of white stones. Katherine has placed some otoliths in little dishes and tells me I can pick them up.

Otoliths are beautiful, pale white and moonlike in color. I lift one whose label identifies it as coming from a grenadier, a big-eyed, deep-sea species that can grow as long as an adult human is tall. I hold the otolith gingerly, convinced it will shatter. Contrary to what the word "stone" implies, it is quite flat, the size and shape of the print left by lipstick from a kiss, though smaller, perhaps. The otoliths grow in the fishes' heads over time, layer by layer like a tree, by accreting minerals such as calcium dissolved in seawater. You can see these layers in the thinner ear-stones. This calcium and protein matrix makes a durable, smooth, cool object, which I find both wondrous and ghoulish. This one I am holding grew slowly in some fish's head out in the vast Pacific.

As I move along the table, the shapes vary: flatter, thicker, scalloped, curved. I think of jellyfish, Space Invaders, lentils. Their size doesn't always correlate to the body size of the species. Big fish can have tiny otoliths, and vice versa.

Tornabene holds up one otolith on its edge, like a coin about to go in a slot. The otolith sits edge-on like this in the fish's head, he says. Fish have three otoliths per ear: The largest stone is called the sagitta, and these tiny vials contain sagitta pairs.

Each set comes from a fish whose species, location, even the vessel that caught it, is recorded on its box: I spot various species of grenadier, saffron cod, rockfish, hake. These are valuable research materials.

After I admire the specimens, Katherine walks me out past a large warehouse, whose interior resembles that at the end of *Raiders of the Lost Ark:* Towering steel shelves line the space, back into the distance, holding stacks of boxes with vials of these opalescent ear stones. Seventeen thousand boxes, each with 140 vials, making up some two and a half million pairs of otoliths.

Because fish accrete otoliths over time, scientists can use these stones for valuable data. Where the fish has been, what it's eaten, and what the water chemistry has been like on its path.

This makes me wonder what all these animals were listening to. For humans, our hearing is centered on, and exquisitely sensitive to, the frequencies of our own voices. That suggests communication was a major driver in the evolution of our hearing range. What about fish? As those first early ears evolved, what influenced their development?

ARTHUR N. POPPER is now retired from the University of Maryland (College Park). Popper has authored numerous papers and edited many tomes on hearing and bioacoustics, and even helped develop international standards on underwater sound measurement.

Popper and his colleague Richard R. Fay have long argued that the sense of hearing initially evolved to allow animals to perceive critical information in the world around them.

"Think about it," Popper says. "When my daughters were young, how did I know things were going on around the house? I *heard* them. You're using sound much more for the world around you than for communication."

Some fish have evolved structures that increase their hearing sensitivity. Many fish have an air-filled sac in their bodies called a swim bladder that helps them control their buoyancy. This sac vibrates in response to sound pressure as sound moves through the fish. The swim bladders of some fish, for example those of cod or some soldierfish, have evolved extensions toward the fish's head that carry vibrations right up next to the ear. Fish might evolve these extensions if there was a benefit to hearing at higher frequencies, with more sensitivity than their ear, which otherwise hears best at a few hundred Hz and tops out below 1,000 Hz.

Other "hearing specialists" are the freshwater group that includes carp, catfish, and goldfish. These fish have evolved a chain of small bones, in fact modified vertebrae, that extend from the swim bladder to the ears, creating a connection.

But not all these fish, specialists or not, make noise at the frequencies to which they are most sensitive. Goldfish do not make sounds at all. So, communication was unlikely to have been a significant pressure on the evolution of their hearing.

Consider also that many invertebrates aren't known to communicate with sound, yet they detect it. This suggests the sense of hearing evolved to expand fishes' and other animals' perception of the space around them, to build a picture of what, who, and where. Though communication would certainly influence the evolution of hearing in species that do make sounds, the sense offers animals so much more.

"Sound," Popper tells me, "gives you a world."

AS SENSES EXPANDED animals' conceptual worlds, their colonization of land was a major turning point in the evolution of life on Earth about 400 million years ago. They were likely looking for food or escaping hungry predators. From dark, cold water, they emerged to light, radiation, and reactive oxygen. They had to bring the sea with them so that certain parts would still work, which they did with aqueous eyes, salty blood, and wet inner ears. As biologist and philosopher Peter Godfrey-Smith wrote, "The chemistry of life is an aquatic chemistry. We can get by on land only by carrying a huge amount of salt water around with us." Inner ears remained fluid-filled, and vertebrates now had to somehow get sound from the air into that wet inner ear. But this is harder than it seems, and the reason is something called *acoustic impedance*.

Crudely put, *impedance* is a measure of how easily a sound wave moves in a given medium, and sound has trouble crossing between mediums whose impedances are quite different. The molecules lose almost all energy, and instead, sound bounces back into the medium it's already moving well in. Basically, the sound hits a wall.

Vertebrate ears didn't change much for about 100 million years after animals evolved beyond the sea and onto land. That might be because the first land animals sensed ground-borne vibrations directly into their inner ear, through jaws that rested on the earth. But at some point, ears evolved a canal from the air that ended with a membrane stretched across it like a drum—the eardrum. This membrane vibrates minutely with sound pressure. Air-adapted ears detect the pressure of a sound wave instead of the motion of the particles.

These same early land animals evolved an amplifying chain of small bones, which were originally part of the animals' jaws. Over millions of years these shrunk into a chamber that became the middle ear. (That these bones are from the jaw is more evidence that early land animals likely detected sound through jaws using ground-borne vibrations.) In mammals, a chain of three tiny bones transmits vibrations from the eardrum to the window of the fluid-filled inner ear, amplifying the energy from the eardrum about twenty times. Due to their appearance, the bones are called the hammer, anvil, and stirrup.

The fossil record indicates that fish evolved into amphibians. Amphibians were the first major vertebrate group to diverge from fish. Some amphibians evolved into reptiles, and some reptiles evolved into birds, others into mammals. Over time, in each group, the patch of inner-ear hair cells got longer. And in mammals the patch had to coil up to fit in animals' cranial cavity without projecting into the brain. In humans, this coiled membrane is now found inside the little spiral in the inner ear called the cochlea. Here, sound becomes a nerve impulse and is sent to the brain.

Ultimately, among the dizzying array of configurations, from corals up to humans, every structure that has evolved to detect sound involves moving the hair on some type of hair cell, and so changing a physical vibration into a nerve impulse.

BRANDON CASPER IS a research physiologist at the U.S. Navy's submarine base in Groton, Connecticut, where he studies underwater sounds and hearing in U.S. Navy divers. I ask him to explain exactly how humans do hear underwater, and he asks me to imagine I'm underwater, facing someone else making a simple sound—say, banging two rocks together. (Of course, I imagine myself back beneath the lake, facing my brother, who is holding a submerged Tonka truck and driving it over the pebbles.)

"When the two rocks come together it creates a disturbance that transmits through the water," he says. "One molecule bumps into the next, bumps into the next, and the wave propagates through the water." So far, so good. Casper continues: "The sound will reach you much faster than it would with the equivalent distance in air."

Once the faster sound gets to my submerged head, he explains, it hits an eardrum with water on one side and air on the other. As the rocks' vibration tries to transfer from water to air, the boundary between these drastically different mediums will reflect most energy away from my eardrum. Not to say the eardrum doesn't move at all, "but the system is just nowhere near as efficient as it would be in air."

Casper has done experiments with navy divers and says that underwater sound probably doesn't reach their cochlea via their eardrum, at least not by much.

"We believe it's via bone conduction," he says. "The skull and our bones are actually very good conductors of sound. Sound hits the skull and then transmits that acoustic energy directly to our inner ear." The amplifying bones and that marvelous eardrum mechanism that we evolved on land? Sound mostly bypasses it all, underwater.

Divers wearing full neoprene hoods hear differently beneath the waves than divers with no hood, whether their ears are covered or not, because the foamy, air-filled neoprene blocks some sound, depending on the frequency, from entering our skulls.

I ask Casper why underwater our voices are so faint. It's impedance. "You've got lungs and your mouth cavity filled with air. And you're trying to push all that air and sound into the much denser medium of water and it just—doesn't."

I remember one more odd effect during that long-ago truck game: That sound seemed to come from everywhere all at once. Casper says that's because of how animals with pressure-sensing ears pinpoint that all-important piece of information from an acoustic signal: *where* it's coming from. When we eschewed particle motion, and its directional data, we needed another way to tell where sound was coming from, so we used the short difference in arrival time between each of our two ears. Sound in air moves at 343 meters per second, yet our brain is fast enough to tell when it strikes one ear earlier than the other, and to triangulate a source from this lag.

"It's called intra-aural time difference," Casper says. "It's what one ear hears versus the other." There's also a slight volume difference. A sound from the right side sounds slightly louder to your right ear than your left, and vice versa, thanks to your skull's 'acoustic shadow.'

"It's that combination of timing and level that we use to ID the source of a sound," Casper explains. It's why humans are good at telling if a noise source lies to their left or to their right—and not so good at whether the sound is coming from the front or back, where the time and volume differences of two sounds are similar.

"All of those cues and techniques that we as terrestrial animals use to detect sound go away underwater," Casper says.

That is why I never gave much thought to sound underwater, why the "silent world" concept took easy root after Cousteau's film, and why Simpson never thought of sound as a meaningful cue for fish, let alone coral.

So how did we start to listen underwater? It only took a few millennia.

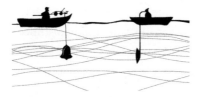

CHAPTER 3

Guns, Quartz, and Arias:
How We Learned to Listen Underwater

The more basic question to answer is not when did men design this or that piece of deep-sea equipment, but what scientific question did they hope to answer by designing it? Why, in other words, have men wanted to study the sea?

—Susan Schlee, *The Edge of an Unfamiliar World*

THE EARLIEST SURVIVING WESTERN WRITINGS about sound underwater are by the Greek philosopher Aristotle. He lived from about 384 to 322 BCE, and he wrote a lot, opining on animals, philosophy, souls, literature, logic, gods, and more. One of his treatises, *De Anima* (On the Soul) from around 350 BCE, contains the passage ". . . sound is heard both in air and in water, though less distinctly in the latter"—something anyone who had ever gone swimming likely already knew.

Aristotle also wrote elsewhere about whales and even fish that made sounds: "the lyra or gurnard, and the sciaena (for those fishes make a grunting kind of noise) and the caprus or boar-fish in the river Achelous, and the chalcis and the cuckoo-fish; for the chalcis makes a sort of piping sound, and the cuckoo-fish makes a sound greatly like the cry of the cuckoo . . ." He even cited the mechanisms of fish sound, rubbing prickly gills or "internal parts about their bellies." Aristotle also averred the sounds were not "voices" but incidental sound, like that of a bird's wing.

Interestingly, Aristotle didn't think animals needed ears to hear. He dubbed fish earless, along with birds and even some marine mammals. "Now, the seal," he allows, "has the passages visible whereby it hears; but the dolphin can hear, but has no ears, nor yet any passages visible."

The Romans also knew from expericne that fish could hear. By the first century BCE, some villas installed *piscinae*—ponds stocked with exotic fish. Some fish were beloved pets who could come when called.

Pliny the Elder was a Roman who lived several hundred years after Aristotle and had his own oeuvre of encyclopedic writings. In his *Natural History* from 77 CE, Pliny describes: "In the Emperor's aquarium, the various kinds of fish come in answer to their names." The triumvir Marcus Licinius Crassus, a Roman general and statesman who was thought to be the richest man in Rome, reportedly sobbed when his pet eel died. One wealthy matriarch allegedly adorned her eel with jewelry. Pliny, like Aristotle, also wrote that however fish were hearing, it was not with *ears,* as they lacked even the tiniest earhole.

Greeks and Romans, then, knew sound worked underwater, and that some aquatic animals heard, but were still largely limited to observations of animal sounds above the waves.

The next so-called scientific record of sound underwater didn't come until 1490 CE, when Renaissance polymath Leonardo da Vinci mused, "If you cause your ship to stop, and place the head of a long tube in the water and place the outer extremity to your ear, you will hear ships at a great distance from you."

In 1552 the anatomist Bartholomeus Eustachius described a critical body part; the spiral cochlea in human ears, clearly the part of the inner ear that *heard*. Renaissance anatomist Julius Casserius was among the first to describe fish ears—sort of. His lavishly illustrated 1601 treatise *De vocis auditusque organis: historia anatomica* contains a dissected fish head showing three semicircular canals: well-known from human anatomy as an inner ear part, but lacking the critical cochlea. Ergo, early anatomists asserted, in contradiction to Aristotle and Pliny, fish *had* ears but could not actually *hear*.

One might well wonder why fish hearing was of interest among the learned set. It likely had to do with larger social forces. In 1543, Copernicus' declaration that Earth revolves around the sun became one of the first pivotal discoveries of the Scientific Revolution, and in 1620, Sir Francis Bacon published his *Novum Organum* (or "New Tool" of logic)—one of the first formal descriptions of what would become the scientific method. (Bacon also made mention of sound working underwater, noting that one could hear gravel grating beneath the water if dragged by a pole.)

It's hard to overstate how profound a change in worldview the Scientific Revolution was in Europe. Traditionally, the Christian worldview decreed human dominion over everything, or, as in the schemes of Aristotle, that humans sat atop a heirarchy of ranked animals. But science upended this. Understanding no longer came from God but from observations and measurements, which transformed the relationship between humans and nature. Now relationships between species could be deduced from anatomy, structures, and behaviors.

In 1735, Carl Linnaeus created the bionomial scientific naming conventions we still use today and placed species in relation to each other. Though Charles Darwin's *On the Origin of Species* would not arrive until 1859, debate on how animals related to humans was now based on observations of anatomy, abilities, and senses. Hearing structures became a point of discussion in part because hard ear bones fossilize and can be easily studied.

Most scientists accepted that dolphins and whales could hear, and assumed ocean invertebrates could not. But fish were puzzling: They lacked a cochlea, but seemed to react to some sounds. If they were hearing, it meant the cochlea was not the only way to hear. Either way, it affected our understanding of the sense of hearing—and how fish fit into the scientifically determined tree of life.

Bacon was in the "fish can hear" camp. In Isaak Walton's 1653 classic *The Compleat Angler*, master fisherman Piscator is cautioning newbie Venator not to startle fish by making sounds. He tells his protégé that although the idea may seem strange, none other than Sir Francis Bacon had shown that sound can carry through water and fish can be startled. But even Bacon's opinion wasn't enough. The debate continued—and it became truly action-packed.

IN 1762, JOHN HUNTER was strolling about a fishpond in a noble friend's Portuguese garden. Hunter was a fellow of the Royal Society of London, which was, at that moment, disputing fish hearing, and Hunter realized he could conduct an impromptu experiment. He asked a gentleman who was with him to hide behind some shrubbery so the fish couldn't see him and fire a gun. He did so. Hunter watched fish flee to the bottom of the pond in a muddy cloud. Fish, Hunter informed the Royal Society, could hear.

But one shotgun blast does not a debate settle. Other scientists tried another tack and removed the tubes of lampreys' inner ears. The lampreys didn't respond differently to sound but they did have trouble orienting. Then in 1896, Austrian physiologist Alois Kriedl, who heard tell that trout at a market pond came to be fed at a bell, put some goldfish in a tank and then rang a bell and blew a whistle beside the tank: nothing.

They *did* react to banging on the tank, clapping, and a pistol shot. He poisoned them lightly with strychnine: Their reactions became more sensitive. He removed their inner ears: no change in their response to sound, but they did seem disoriented. Kriedl concluded that fish did not hear: If they did sense sound, it was with some other body part, perhaps some sense on their skin.

In hindsight, it's easy to see how these observations would be confusing, since fish were almost certainly reacting to sound with their lateral line as well as their ears.

In the early twentieth century, scientists subjected fish to car noise, whistles, knocking, annoying children's toys—all with ambiguous results. Beginning in 1909, Harvard biologist George Parker did a series of experiments on dogfish. He swung a pendulum at the wooden walls of a fish tank, measuring if the dogfish shark inside reacted. He cut the nerves connecting ear and lateral line to the brain; and even numbed the animal's skin with cocaine, in case the skin felt vibrations. The shark with intact ears responded best. Having eliminated every other cue, Parker concluded that sharks could hear.

German ethologist Karl von Frisch is perhaps best known for discovering how bees' flight pattern, also called a waggle dance, communicated the location of food to the hive. Years before this work, in 1921, Von Frisch came to work at Rostock University, on the Baltic Sea, where he became interested in fish hearing. Rostock's ear clinic was directed by one Professor Otto Körner, a researcher who studied catfish and was firmly in the "fish are deaf" camp. Frisch's interest was piqued: "The fact that a principle was involved explains the heat engendered by this controversy," he wrote. The principle in question, as it had been for so long, was whether fish could hear without a cochlea, and so prove the little spiral was not the only way to do so.

Körner had exposed his catfish to various sounds, even summoning a famous soprano to serenade the tank. Presumably, the fish would respond, but nada. Frisch thought this was short-sighted. If he were a fish, he might not respond either. It was the kind of oversight that a good behaviorist must account for: An animal doesn't react to *every* stimulus, just the ones that relate to its ongoing survival. Opera wasn't relevant to a fish. Food was.

Frisch got his own catfish, christened him Xaverl, and took out his eyes. He put a hollow candlestick in the tank as an ersatz home for "the little blind fellow." Each day for several days, he lowered a piece of meat into Xaverl's tank while ringing a bell. The fish smelled the meat and swam out to eat. After six days, Frisch rang the bell sans meat. Xaverl, with no cue but the sound, swam out of his candlestick to eat. Frisch invited Professor Körner to see.

"While the kindly old gentleman sat in front of the aquarium waiting for the experiment he was convinced could not succeed, I went to the farthest corner of the room and gave a soft whistle. Xaverl at once came out of his tube, and at the same time the Professor, almost visibly deflated, could be heard to say reluctantly: 'There is no doubt. The fish comes when you whistle.'"

Frisch published his work in 1936, and his students at Munich's Zoological Institute took up fish training. To avoid cross-contaminating training cues, students scattered their tanks in various departments: Signs throughout the building exhorted people not to whistle or sing, please, lest the sounds confuse the fish.

Frisch never conclusively proved *why* fish heard, but figured it was for the same reason any other animal sensed things: to find food, to pinpoint danger, to help them make choices. All the things "which mean something to a fish though they are concealed from the tourist, who contemplates the stillness of a sheet of water."

Meanwhile, only a handful of scientists were asking about the hearing of other underwater animals, with only scattered observations of how octopuses, crabs, and shellfish seemed to react to sound. Naturalists guessed as early as 1910 that octopuses might respond to sound waves, likely because they were one of the few species often kept in aquariums, where people noticed such things. So, by the eve of World War II, it was long known that mammals could hear, fish hearing was settled, and only a few questions had been asked about invertebrates.

POST-ENLIGHTENMENT, THE DISCIPLINES of biology were studying animals' hearing. Meanwhile, the physicists were asking a question un-related to animals: Does sound travel through water, and if so, how fast? It was inquiries about sound speed that would catalyze the technology that let us listen to the sea.

Mathematician Pierre Simon, Marquis de Laplace (1749–1820) wrote equations relating the speed of sound in a medium to density and compressibility. But that meant a material must be compressible to carry a sound—and water was not thought to be. Sound speed was thus linked to questions of water's fundamental properties.

In 1743, French physicist Jean-François Nollet dunked his head into the Seine in Paris, while above the water a loud voice, a bell, a whistle, and a gunshot all sounded off. (He concluded sound worked underwater, therefore water was compressible.) This, perhaps, was the first "scientific" experiment on sound underwater.

Scottish polymath Alexander Monro Secundus tried to measure sound's speed in water in 1785. For his experimental design, he seems to have taken inspiration from thunderstorms. Anyone who's watched lightning strike and heard thunder rumble knows light travels nearly instantaneously but sound travels slower. Monro figured he would mea-sure the difference in travel time between a light in air and a sound in water over some distance, and so calculate the speed of sound.

So Monro went to a lake—at night—and lay with his ears sub-merged but his eyes above the surface. An assistant 800 meters away simultaneously let off a loud explosive in a submerged bottle and fired a gun with a bright powder flash. The charge and the flash appeared simultaneous to Monro. But that was because 800 meters wasn't far enough to reveal the lag between sound and light.

It was hard to find a body of water large enough for a meaningful experiment. In the early 1800s, French geologist François Sulpice Beudant went to the sea near Marseilles to try but the split-second timing was challenging. He estimated the speed was about 1,500 meters per second. Then young Swiss physicist Jean-Daniel Colladon read Beudant's book and reached out to him to ask about the details of his experiments.

Colladon then designed an experiment on Switzerland's Lake Geneva, which was 14 kilometers wide and more accessible than the ocean. In 1825 he tested bright light sources (including a rocket that badly burned his hand) and decided on gunpowder flash, and underwater noisemakers including striking a submerged anvil. The next year he tried better tools, such as a sunken bell and hammer. But he could hear even the loudest sounds from about 2 kilometers at most, and he wanted the experiment to span the whole lake to make the time lag obvious. He had to get louder or listen better. Then he remembered reading about da Vinci's tube.

After trialing listening devices including a spouted watering can, Colladon built a three-meter tin tube with a large horn on one end. (Colladon's middle ear, and the tube, were still full of air, but with his ear against the tube, some sound probably traveled through bone to his inner ear.) Colladon enlisted his father as an assistant and onto the nighttime lake they drifted in their respective boats. (They even measured the lake's temperature: 8 degrees Centigrade.) Colladon had built a Rube-Goldbergian dream of levers to simultaneously ignite the powder and hammer the underwater bell, without relying on human reflexes. And as he floated near the town of Thonon, through his ear trumpet he heard the bell near Rolle, 14 kilometers across the lake.

Colladon had meanwhile partnered with French mathematician Charles Sturm who had worked in the laboratory on supporting measurements. In 1827, Colladon and Sturm reported the speed of sound in water as 1,435 meters per second, nearly four and a half times faster than in air.

Each medium has a constant speed of sound: It bumps through air of 0 degrees Centigrade at about 331.5 meters per second, about 343 meters per second under room-temperature conditions. Water, much denser than air, conducts sound four and a half times faster. (In compressible solids like steel molecules are packed tighter. Sound races at about 5 kilometers per second through stainless.) A sound wave also loses energy less quickly in water than in air. So, underwater sound moves faster, goes farther, and loses less energy along the way.

In their report Colladon mused that sound from a sufficiently loud bell might travel 60 or 80 kilometers. It might also be possible, he wrote, "to use the echo from the bed of deep oceans to measure depths . . ." With these words, he voiced the possibilities that would lead to sonar. But it would be a long path to get there.

OUT IN THE nineteenth-century sea lanes steamships had arrived, commerce was booming especially in the newly formed United States, and mariners needed accurate naval charts. No merchant wanted to lose cargo to fog or hazardous shoals. Imperial powers were funding big, multiyear research expeditions, (such as those by Captain Cook, whose first voyage was to observe an astronomical event, the 1769 transit of Venus, from Tahiti, and discover whether there was a large land mass south of Australia). As such endeavors searched out and claimed foreign shores, they needed maps. And naval warfare was ever-present. A country's maritime power required maritime data.

The U.S. government created the U.S. Coast Survey in the early nineteenth century and, a few decades later, the Depot of Charts and Instruments. Surveyors mapped coastlines and currents but to measure depth they had only sounding lines. These long cables reeled out with weighted "sounders"—bell-shaped cups that brought up samples of the seabed so that geologists could characterize the composition of the seafloor: mud? sand? rock? Sounding was time consuming, and one reading could take an hour. During one three-and-a-half-year voyage in the 1870s a survey ship took only three hundred soundings in water deeper than 1,000 fathoms. (A fathom is 6 feet, just under 2 meters, so these readings were in the deep sea some 2 kilometers down.)

Following Colladon and Sturm's 1827 report Charles Bonnycastle and Robert M. Patterson wondered if, as a subsequent report described, "an audible echo might be returned from the bottom of the sea, and the depth be thus ascertained from the known velocity of sound in water." On August 22, 1838, Bonnycastle and Patterson set out aboard the Coast Survey brig USS *Washington* armed with a modified stovepipe and the standard noisemakers: a pistol, a bell, and a softball-sized explosive charge called a petard. Bonnycastle sat in a boat 400 meters away from the ship with a 2.5-meter tin listening tube. He strained to hear a submerged ship's bell being struck, particularly its echo. But the sea's vigorous sloshing on the tube drowned out the bell, and any echo.

A few days later they tried again. Through his tube 137 meters away Bonnycastle heard two sounds, one-third of a second apart. Was this the echo from the seabed? That time lag would mean the depth was about 300 meters. To confirm, the crew took a sounding with an old-fashioned line, which then spooled out more than a kilometer beneath them. Either echo-sounding was impossible, Bonnycastle concluded, or "some more effectual means of producing it must be employed."

In 1859, Lt. Matthew Fontaine Maury, director of the Depot of Charts and Instruments, repeated Bonnycastle and Patterson's experiment. He failed. Tube or no tube, as long as humans were listening with ears, echo-sounding failed. The first subsea telegraph cables were laid in the 1850s, and their surveyors surveyed the seafloor contours, water depths, and landing sites with sounding lines.

There was, however, interest in using underwater sound to make coasts safer. Engineer Arthur Mundy proposed that lighthouses or lightships send out underwater warning tones to ships at sea, and in 1899 he and inventor Elisha Gray patented an underwater electric bell and microphone, a device that would evolve into a "hydrophone." In 1901, Mundy and others formed the Boston-based Submarine Signal Company (SSC). Their early hydrophones overcame the air/water boundary by being submerged in water tanks bolted inside ship hulls. But the ocean was still *loud*. Wind and waves sometimes drowned out the bells.

Then, tragedy accelerated innovation when in 1912, an iceberg sent the "unsinkable" *Titanic* to the bed of the Atlantic Ocean just off the Grand Banks of Newfoundland.

REGINALD FESSENDEN STOOD on the U.S. revenue cutter *Miami*, above Newfoundland's Grand Banks, on a chilly April day in 1914 and watched the iceberg loom closer, thirteen stories above the chop. The berg dwarfed *Miami* as Fessenden guided a device that looked like a large snare drum off the ship's side. This was an oscillator, which could both make and detect sounds underwater, combining the functions of a bell and microphone in one device. Oscillators were already used in telegraph technology, and that day Fessenden would test if this underwater version could detect objects, specifically icebergs, by making a sound and then detecting its returning echoes, for these very waters had claimed *Titanic* two years earlier.

Titanic's foundering had brought a flurry of patents for iceberg-avoiding technology. One early proposal by Sir Hiram Maxim was for the "collision-preventing apparatus"—a device modeled after bats' "sixth sense" (it wasn't yet called sonar). At that time, people thought bats used sound from their wings to navigate, and Maxim proposed sending out low-frequency sound from ships and listening for echoes from bergs, but the device was meant for use in air. Fessenden, a telegraph engineer, wondered about doing this below the surface. After all, an iceberg's bulk is underwater. Fessenden built a prototype and asked the government for a ship to test it on the notorious Grand Banks, and the U.S. Navy's *Miami* set out into fog and rough weather. Fessenden was unable to sleep for being thrown out of bed by the waves.

As *Miami* stopped 150 meters from the iceberg Fessenden lowered the oscillator and sent out its sound. And there it was—an echo that seemed to come from the correct distance.

But 150 meters was too close to a berg to be a useful warning for a ship. And was the echo from the berg or the seabed? The captain maneuvered *Miami* farther from the ice. One mile. Two. Two and a half. At each distance, Fessenden heard several echoes. One remained constant, likely returning from the seafloor. But another echo took successively longer to return as the ship moved away. If sounding the depth was "echo-sounding" this was "echo-ranging" or determining the distance from one object to another object.

Nearly two years to the day after the sinking of the *Titanic*, Fessenden had stepped beyond passively listening for signals, and actively used underwater sound to "see" silent maritime hazards—and the bottom of the ocean. Little did he know that later the same year, an unimaginable new hazard would emerge.

IN 1915, RUSSIAN electrical engineer Constantin Chilowski was laid up with tuberculosis at a sanatorium in Davos, Switzerland, chewing over the so-called submarine problem. The Great War had begun and German *Unterseeboots* were creeping through the seas, sinking cruisers with ease from the war's first months. U-boats could not be seen, smelled, or touched. The only way to detect them was to listen for their growling engines. And this was the problem.

British ships were outfitted with hydrophones, but they were clunky. The listening vessel had to be powered off to hear anything, and even then their range was limited. Plus it was difficult to tell which *direction* a sound source lay in.

Detecting submarines by actively using sound was tantalizing. And Chilowski knew existing devices, like Fessenden's oscillator, could actively "echo-range" an iceberg or the seafloor. But such devices made sound in all directions, and so the echo still didn't pinpoint the sound source.

It was possible to focus sound, like any other wave, into a beam. A higher-frequency sound means a shorter wavelength: lower, longer. Because the wavelength of sounds made by existing oscillators was relatively long—even more so in water than air, thanks to sound's greater speed—focusing the sound required an unreasonably huge apparatus. But a high frequency sound would have a shorter, more feasibly focused wavelength.

A higher frequency sound would also have a better reflection off objects smaller than seafloors and icebergs. As objects get smaller, the wavelength required to echo well off of them get shorter. If the wavelength is much smaller than the object, you get good reflection. If it's the other way around, you get poor reflection. It's like waves on a lake reflecting off an island (yes): a boulder (some); or a reed (no). A German U-boat of the time was about 60 meters long, and in cross-section only 6 by 4 meters. Existing oscillator frequencies were low enough that reflections were imprecise.

Chilowski designed and proposed an oscillator with higher-frequency sound. His idea reached the French physicist Paul Langevin, who was also thinking about the submarine problem. Chilowski went to Paris. The duo tested a tweaked oscillator in the Seine River and patented it in 1916.

But the device wasn't perfect. It made the sound, but detecting the echoes was unreliable. Chilowski and Langevin parted ways: Langevin kept working. Then, in February 1917, he had an idea. He'd once worked under Pierre Curie, who had co-discovered something called the piezoelectric effect (*piezen* = Greek for "squeeze" or "push"). Some natural crystals, such as quartz, generated an electric current when they were pressed. Langevin now wondered if very thin quartz crystals might vibrate in response to sound waves' pressures and generate a current—and thereby make for an incredibly sensitive receiver.

Enter Parisian optical craftsman Ivan Werlein, who happened to have a foot-long quartz specimen displayed in his shop window. Langevin persuaded the optician to slice this remarkable specimen into a (very) thin sheet, 10 centimeters by 10 centimeters. And it worked.

Quartz became critical to the war effort. Opticians and jewelers found their wares in demand as collaborating French and British scientists bought up all they could find; according to one account, Canadian scientist Robert Boyle, who was leading the British submarine-detecting efforts, commandeered a warehouse of quartz crystals from a French chandelier-maker. By early 1918, high-frequency quartz oscillators could send a signal 8 kilometers, and echo-range objects 500 meters away—1,500 meters by the summer of that same year. The British called these oscillators anti-submarine devices, or ASDs, and their study "ASDICS." They shared it with Americans in October 1918, just before the war ended.

By the 1920s, the SSC sold an echo-sounding "fathometer," and the first submarine cables to use echo sounding were laid across the Mediterranean. One vessel took nine hundred deep soundings on one Atlantic crossing, a huge improvement over sounding lines. Military scientists scrambled to refine echo-ranging and underwater sound signaling, the technology that would by World War II come to be called sound navigation and ranging, or SONAR.

And as people started sending sound over larger underwater distances, strange new phenomena emerged.

Seawater is not uniform, after all. It's a swirl of water masses, of temperatures, salt concentrations, and other properties that can affect water density just enough to reflect and bend the paths of sound. Untangling, predicting, and leveraging these paths would eventually lead to that underwater secret listening network. One of the first major breakthroughs came when oceanographers and sonar technicians had to solve the "afternoon problem."

IN 1936, THE chief of naval operations in Washington, D.C., wrote to the chief of the Bureau of Engineering requesting permission for *Atlantis,* the scientific ship used by Woods Hole Oceanographic Institution (WHOI) in Cape Cod, to join the navy submarine USS *Semmes* at their base in Guantánamo Bay, Cuba. They wanted the scientists' expertise "in connection with underwater sound investigations of seawater conditions." A weird problem had arisen: As seamen honed their sonar, they found it worked differently in the morning than in the afternoon.

The sailors would head to sea at dawn and their sound beams hit their targets perfectly: by midday, they missed. At first, operators blamed human error and even the men's post-lunch laziness. But eventually they realized it must be the ocean itself. So they turned to ocean scientists.

The WHOI scientists suspected some daily phenomenon linked to sunlight and how it could change the temperature of the water over the course of a day. One scientist wrote, "There is something extremely interesting there from the oceanographic standpoint, in the regularity [note in pen: "or the reverse"] with which solar energy penetrates the water."

By August, *Atlantis* and *Semmes* were sending out sound beams throughout the day while tracking the water's temperature at various depths. At dawn, the seawater was mostly uniform from the surface downward, but as the day wore on, the sun gradually warmed the upper layers by a degree or two.

We know that when a sound wave meets a different medium, some of it reflects. But when it meets the new medium at an angle, the sound may also *refract*—passing into the new medium at a different angle. It bends toward the medium with the lower speed. And this speed can change with even small shifts in a medium's density.

Three factors can affect seawater density: temperature, pressure, and salinity. Of these, temperature has the most dramatic effect. Between freezing and near-boiling temperatures, the warmer the water, the faster sound travels through it. So with the afternoon problem, different water temperatures were making water layers behave like different mediums, refracting and therefore bending the sound wave down, ever more as the day went on and the top layer warmed.

With the afternoon problem solved sonar operators could tweak their beams to compensate for the temperature change and keep listening all day. They had illustrated something vital: Where water isn't uniform, sound bends and curves.

WORLD WAR II erupted in 1939. Sonar technicians now measured the temperature at various depths and then, with these "sound speed profiles," drew the predicted path of their sonar signals in large graphs called ray tracings or ray diagrams, lines arcing sinuously through the

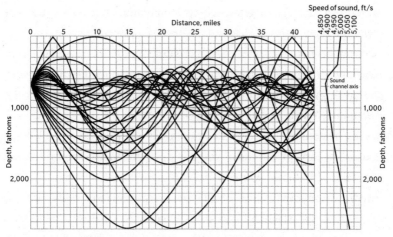

Adapted from Robert Urick's textbook *Principles of Underwater Sound* (New York: McGraw-Hill, 1973), p. 160, Figure 6.9.

water. These tracings revealed phenomena of military significance, including *very* long-distance sound transmission.

In most oceans, the deeper you go, temperature *decreases*, quickly at first and then slower; pressure *increases* at a regular rate; and salinity (usually) *increases* because dense salt sinks. That means depending on where you are you can usually make educated guesses about how the water layers are configured around you, and thus how sound will bend in them.

In a deep mid-latitude ocean, off, say, Massachusetts or Bamfield, surface water is warmish. Deeper down, the temperature drops and then remains constant. Pressure and salt effects take over, and so density—and sound speed—rises again until you reach the seabed. This makes a layer where the sound speed reaches a *minimum*. In warm latitudes, this layer is generally a kilometer or more down. In cooler water, it's shallower. In the Arctic, it's at the surface. Since sound refracts toward the lowest speed, sounds in this layer continually bend back into it, and a sound wave can travel uninterrupted for hundreds, even thousands, of kilometers.

In 1941, geophysicist Maurice Ewing of Lehigh University joined the WHOI and *Semmes* team. Ewing was among the first to realize this ocean layer could transmit sound farther than anyone guessed. In 1945, he tested his theory. Off the Bahamas, he dropped TNT into the channel. A hydrophone waited 3,000 kilometers away, off Dakar, Senegal. The charge was dropped; and the explosion came through above the Atlantic's ambient roar.

The especially sound-conductive layer was dubbed the SOFAR channel—short for SOund Fixing And Range, named for a proposed navy scheme to locate fighter pilots shot down over the ocean. The system didn't come to fruition, but the term "sofar channel" stuck.

In 1946 the newly minted U.S. Office of Naval Research decided to leverage that amazing ocean layer in which sound could travel so far, and as the specter of the Cold War loomed, build an array of secret hydrophones in the channel that could listen—all over the world.

AT&T built the first hydrophone station at Eleuthera in the Bahamas in 1952. Other stations followed along the U.S. East Coast, the stations' construction often explained under the guise of "oceanographic research." The stations consisted of offshore hydrophones cabled to shore stations where data spilled out on long paper coils. Technicians studied the readouts for anything suspicious. By the 1970s, listening stations dotted the Pacific coast, Iceland, Hawaii, Guam, and Wales, where thousands of personnel listened. The network was dubbed SOSUS: the Sound Observation and Surveillance Undersea Network. And it was kept secret.

Wartime listening from World War II through the Cold War opened up the ocean to our ears, and to biological sounds no one had dreamed might exist.

IN THE WINTER of 1942, to listen for enemy submarines gliding up Chesapeake Bay to the nation's capital, a hydrophone array was built across the bay's entrance. In May the listening stations were suddenly beset by "unaccountable noise troubles." It sounded like a frog chorus punctuated by rapid drumming. The sound also cycled, rising in the evenings and overnight.

Columbia University scientists tested if the sounds were biological. One night they exploded five blasting caps right beside the hydrophone. They reported, "Instantly the cacophony was hushed—exactly as if a gun had been fired in a cage full of chattering monkeys!" Something living had been scared, a conclusion strengthened when the noises started again moments later.

The navy asked local fishermen if anything they knew made such a noise and learned that one local springtime catch was the "croaker," a 30-centimeter-long fish that sometimes made an eponymous croak when it was hauled out of the water. Croakers came to the Chesapeake each spring to breed, and males made a loud hum to attract females. Isolating a croaker with a hydrophone, the source of the sound was confirmed.

At the time only people who spent a lot of time around fish, such as fishermen, might hear their sounds. The broader public assumption— still—was that most ocean animals beyond, perhaps, a few whales, were silent.

But on the military hydrophones, astounding new voices drifted from the deep, regular pulses, knocks, and blips. The reports show that while listeners had no idea which animals were making them, they became familiar with certain sounds.

One common sound, a string of steady knocks, was known among sonar operators as "carpenter noises" or "carpenter fish."

Another mystery sound was a 20-Hz pulse, sometimes called "BLIPS." This was an extremely low-pitched sound, inaudible to all but the very best human ears, but clearly visible on the hydrophone printouts. It resembled a slow and steady heartbeat—*blip-BLIP, blip-BLIP.*

One sound was so common that it became downright annoying. Submariners noted a ubiquitous, sometimes deafening sound described as frying bacon, rain on a tin roof, or simply a crackling. Perhaps oddest, echo-sounding pulses often bounced off a layer hovering several hundred meters down, a "false bottom," in places where the sea was known to be much deeper. The University of California's War Division mapped this layer from Cabo San Lucas all the way up to Cape Mendocino in California.

Warfare had moved underwater, and these "biologicals" had to be better understood. So, the first systematic studies of animals and underwater sound were done to inform sonar training manuals.

Martin Johnson was a South Dakotan who had come to California to study plankton, very small, drifting creatures that encompass anything that drifts, such as phytoplankton, algae, zooplankton, and larvae. Johnson was a zoologist at the Scripps Institution of Oceanography when the navy hired him to help solve the mystery.

Johnson knew that plankton migrated en masse up and down in the water through the day, following light levels. Could this be creating the "false bottom" phenomenon? In 1945 Johnson motored about 32 kilometers offshore and sent out echoes for twenty-four hours, tracking the false bottom until noon the following day. The phantom seabed rose in the early evening and then stopped several hundred meters down. It diffused throughout the night, sending back a weaker echo. As the sun rose the layer coalesced and dropped again. Their rate, about 2 meters per minute, matched the swimming speed of many plankton. Johnson sunk slim nets down into the layer to confirm. Retrieved, the nets teemed with plankton. This created a buffet for squid and other animals. The sonar was bouncing off the sheer biomass of the layer.

Johnson also surveyed what animals were present where the "crackle" was loudest and pinpointed the culprit. "Snapping shrimp" are thumb-sized shrimp that live on the rocky or sandy bottoms on tropical and subtropical coasts around the world. One claw is oversized and very powerful. The shrimp can snap it closed at 97 kilometers per hour, pressurizing the water inside enough to raise its temperature to 4,000 degrees Centigrade. This flash-boils the water, shooting out a bubble that pops loud enough that it scares the shrimp's enemies and stuns its prey.

Nearly a decade before Cousteau's *Silent World*, Johnson wrote in a 1947 report, "The sea has long been looked upon as a realm of silence. That this view is no longer tenable has been abundantly demonstrated as a result of extensive investigations with modern underwater sound detecting gear."

THE MAP'S CREASES have been flattened behind glass in the New Bedford Whaling Museum, dun-colored land and white water describing the place where the deep, narrow Saguenay River meets the St. Lawrence in Quebec, Canada. The depth contours are interrupted here and there with circled pencil notes: T1, T2. A hasty pen has scrawled a name in the top corner: Schevill.

William Schevill had been a paleontologist until World War II, when he worked with the U.S. Navy and WHOI. He became fascinated by the sea. When the war ended and the navy funded efforts to catalog underwater sounds, Schevill and his wife Barbara Lawrence answered the call. Lawrence was a distinguished scientist in her own right, the curator of mammals at Harvard's Museum of Comparative Zoology who had traveled to Sumatra and the Philippines in the 1930s to study bats.

Schevill and Lawrence faced the same problem as other pioneers like Johnson: If you can't see what animal is making previously unheard sounds, you can't match the sound to a species. The solution was to travel where you know exactly who is present, and thus in 1949 Schevill and Lawrence traveled to see the beluga whales of the St. Lawrence River.

We think of belugas as an Arctic species, but a genetically distinct population has remained in the St. Lawrence Estuary since the last ice age. They ply the brackish waters where the river shades to the sea, just where the Saguenay River's deep, powerful current also rushes from its fjord to mix in. These rich waters near Tadoussac and Baie-Saint-Paul are home to several hundred lumpen white whales.

Two Tadoussac locals went out with Schevill and Lawrence and remarked that they'd heard some of the whales' sounds resonating through their canoe in the past, but never the variety now coming through the hydrophone. This was the first recording of wild whale sounds, issued in 1950.

Schevill and Lawrence kept listening and several years later, Schevill and colleagues also solved the case of the so-called carpenter fish.

Schevill had some clues, because similar underwater sounds had been mentioned before, in a very niche industry. A former New Bedford whaler, Henry Mandley Jr., had told Schevill of hearing impulsive noises through his boat's hull while he'd hunted sperm whales. And an 1840 sperm whaler's account describes a "creaking" sound near the whales. But the whalers had never really connected the sounds to the whales.

In 1957, Schevill and his colleague took to the sea off North Carolina, crept near a sperm whale pod, and lowered a hydrophone. They heard three different sounds: a percussive sequence of thuds; a "grating sort of groan" like a rusty hinge creaking; and clicks. The last was the most common sound, and "loud enough to blacken the sounding recorder paper." The sperm whale, the species to which Moby-Dick belonged, was unmasked as the "carpenter fish." (What these sounds actually meant among sperm whales would take a while longer to determine.)

In 1963, during the height of the Cold War, Schevill, William Watkins, and Richard Backus also put to rest the mystery of that low double-time blip that thudded through the seas. They localized the sound to fin whales. Fin whales are the second-largest whale in the world after the blue whale, and they have perhaps the most regular and monotonous voice in the ocean. A fin whale has only a few calls, all low-frequency, but it makes them a lot and the most common is centered at 20 Hz.

Schevill died in 2004. According to an obituary written by W. D. Ian Rolfe, "Bill helped defuse a tense moment between the USA and Soviet Union during the Cold War. The US military suspected that low frequency blips were being used by the Soviets to locate American submarines, whereas Bill showed these were produced by fin whales (*Balaenoptera physalus*) hunting prey."

THE APTLY NAMED Marie Poland Fish had gained scientific renown when she solved the long-standing mystery of where American eels breed (the Sargasso Sea). In 1936, she started a marine lab at the University of Rhode Island and subsequently became state ichthyologist (state fish scientist). After WWII, when the U.S. Navy started cataloging biological sounds, they tapped Fish to "audition" as many fish species as possible.

Fish visited her subjects in captivity or traveled along the coast to sample them, eventually fencing off a pool near the University of Rhode Island. Her next challenge was getting the fish to talk. Many did so only in certain circumstances, or at certain times of the day or year. But she got creative, and, honestly, it's difficult to read the details of many of her experiments, which include electric shocks, cattle prods, and what might politely be termed "rough handling." She worked in secret, collaborating only with a select few. They recorded some sounds in water, but many in air, with the fish held in the hand. Each "audition" went on tape, with descriptions of "grunts," "growls," "hums," and "barks." (Fish also describes a few invertebrate sounds, from snapping shrimp to the hushed "snap" of byssal threads that anchor a mussel to a rock.)

Not every fish makes sounds, but Fish described more than 150 soniferous fish from Canada to Brazil. And there was an astonishing variety of sounds to describe. Her research, and that of those who followed her, showed that fish have the widest variety of sound-making structures of any vertebrate group. Some scrape and stridulate: Sculpins move their pectoral girdle. Toadfish, squirrelfish, and others drum on their swim bladder with special muscles or tendons, making resonant hums, moans, and boops. Blue grunt or beau-gregory scrape or grind special teeth in their throats. Some fish burp or expel gas from their anus. Herring are inveterate farters. Stunningly, the sound-making mechanisms Fish described—drumming swim bladder, grinding teeth—were quite like those described by Aristotle more than two thousand years prior.

IN A DIFFERENT WORLD, Marie Fish's discoveries might have spurred even more research by marine biologists. But her data was meant to inform military listening, and many of the hydrophones capable of listening to broad swaths of the sea also happened to be the only military tools that could listen for enemy submarines with the onset of the Cold War.

So, the network's capabilities, and the new worlds it revealed, were kept largely secret as Soviet and American subs stalked each other through the seas. And only select marine biologists, like Schevill and Lawrence, and Fish, were privy to their acoustic data.

It's strange to think of now, but in the 1950s and '60s, there simply wasn't a field of study for animal sounds in the ocean, or at least, not one that focused on much beyond whales. Instead, only a handful of scientists were studying topics like what fish heard, or how they made sound. Often, scientists who worked in related but distant disciplines such as human hearing, or oceanography, independently made discoveries, often unbeknownst to each other, that would prove relevant to this infant field. One such discovery around this time was exactly how the cochlea, that little spiral in mammals' ears, worked.

Physiologists had known for centuries that the cochlea in mammals turned sound into a nerve impulse, but not *exactly* how. Georg von Békésy won the Nobel Prize in 1961 for his discovery that the coiled membrane that lay inside the spiral resonated to different frequencies at different points along its length. Lower frequencies vibrated the stiff, narrow end, while higher frequencies vibrated the wider, more flexible end. Sound traveled up the membrane like ripples up a shaken bedsheet, flexing the most, and thus bending the hair cells' cilia most forcefully, at a given point that corresponded to the sound's frequency. The neurons connected to this spot then fired, and so told the brain the frequency, depending on where along the spiral the impulse originated. This sophisticated transduction is called place-coding.

Elsewhere, a very few civilian scientists happened to be researching underwater animal hearing. In 1963, eminent marine biologist Sven Djikgraaf showed that cuttlefish, a mollusk relative, responded to low sounds of 180 Hz, some of the first data on invertebrates and sound. But such endeavors were often scattered among disciplines and scientists who worked in relative isolation.

That would change soon. In 1963, William Tavolga saw the study of underwater animal sounds gaining some momentum. Even outside of the military's detailed data, scientists were realizing that there was a lot to learn. Working in close quarters with animals at newly popularized aquariums revealed how complex marine mammal sounds were, and people like Tavolga himself were already studying fish acoustics. But because they hailed from such different backgrounds, scientists often didn't know about one another's work, and thereby lost opportunities for collaboration. Furthermore, most were focusing on mammals like dolphins. But through his fish studies, Tavolga knew there was much more to explore. And just because the navy's hydrophones were secret, it didn't mean that the study of underwater animal sounds had to be. So, in 1963, he organized a conference in the Bahamas on "marine bioacoustics." It was the first use of the term, and along with another pivotal conference later that year focused on marine mammals, marked the birth of the nascent discipline. It was still niche, but scientists began talking, discovering linkages in their work and new questions to explore.

Tavolga, meanwhile, continued to study fish hearing. One of his landmark studies was done right in the Bahamas with colleague Jerry Wodinsky. The duo gave nine local fish hearing tests. If you're wondering how one gives a fish a hearing test, though a fish can't tell you it hears something, it can be trained to react when it hears a sound. Tavolga and Wodinsky played the fish tones of a given frequency, quieter and quieter, until the fish didn't react. Then they turned the volume back up until the fish reacted again. When the fish reacted about 50 percent of the time, this sound was deemed the edge of the quietest sound the fish could hear. They repeated this multiple times. Testing at different frequencies makes a graph with a U-shaped line called a hearing "curve," or an audiogram. It showed what frequencies the fish heard best which in turn suggested what fish found important. Human audiograms, for example, are most sensitive to the frequencies of human voices. Tavolga and Wodinsky found that fish heard best at around 300–500 Hz, and topped out by 1,000–1,200 Hz. These surprisingly low ranges could pick up the low-pitched grunts and growls of their own species, though they weren't very sensitive to ambient wind and rain, which tend to be higher in frequency. It also shows that an otolith-based ear will tend to hear lower sounds. The range also hinted the fishes' lateral line helped them hear the deepest tones.

(Naturalists take note: On this field season, Wodinsky insisted they take all their notes in pencil, not pen; Tavolga thought this was odd, but he capitulated. Then, on the flight back home, two bottles of Beefeater gin [which were cheap in the Bahamas] broke in Tavolga's suitcase and soaked the study notebooks. The ink graphs smeared beyond all hope: The pencil data was unaffected. After hearing this story, I ensured all field reporting for this book was done in pencil.)

Back at his lab in New York Tavolga built an automatic fish-hearing test tank: the "audio-ichthyotron," where he automated tones, and gave mild training shocks using electronics. The device tested the hearing of several species, systematically. Each fish could take weeks to train, and then days to get the data. Why did certain species hear certain frequencies? Did their sensitivities change over time? Questions unspooled. And one of Tavolga's prospective students began a lifelong deep dive.

An undergraduate Arthur Popper was walking to the Bronx campus of New York University one day around 1964 when he passed a pet-store display he hadn't noticed before. He had time to kill before his bus came, so he wandered in. There, swimming amid the tropical fish, he noticed a fish with no eyes. This, the proprietor explained, was a Mexican blind cave fish. It lived in pitch-black flooded caves in Mexico and had lost the ability to see.

Popper was fascinated. How did it get around? What was its world like? He described the fish to his undergraduate advisor, anatomist Doug Webster (who, incidentally, studied hearing, but in desert rodents). Webster arranged for Popper to do a project on the cave fish, and he dove into study. Eventually he learned that a researcher at the American Museum of Natural History, William Tavolga, also worked on this curious topic of fish hearing.

Though there was, technically, a freight elevator, "the more interesting way up to his lab was a spiral staircase," Popper says. "In the middle of it was a platform, with a huge stuffed Galápagos turtle. It was a great place. I remember it with infinite fondness."

The two met and Tavolga agreed to take Popper as a student if he came up with his own project. Popper read through Tavolga's papers and, just before their next meeting, came across the sentence "No one has ever figured out if fish can do sound source localization." That means, basically, telling where a sound is coming from. It's one of the most fundamental functions of hearing because the sense evolved to help animals locate things at a distance. Popper had found his project.

He set to work. He published the cave-fish audiograms in 1970, after which he dove deeper into the patterns and principles of fish hearing. Newly married, he went to the University of Hawaii to teach and do research on fish hearing in 1969 after finishing graduate school. And here he met several significant researchers.

Popper had once met Georg von Békésy after a talk and asked him what he was working on. The laureate replied, "Fish hearing," which had depressed Popper at the time. What could he contribute, if a Nobel Prize winner was getting involved? Now, in Hawaii, he met Békésy again at their faculty apartment building garbage dump: He turned out to be a neighbor. Popper asked him how the fish-hearing research was going, and Békésy said he'd given it up, as it was too hard. Which, Popper says, plunged him into a different sort of despair.

In December 1971, at a barbecue, Popper also met Békésy's postdoctorate Richard Fay, who had also studied fish hearing as a graduate student at Princeton University. Fay would eventually become one of Popper's lifelong collaborators and closest friends, as the two dove into the nuances of fish hearing.

One question Popper remained deeply interested in was sound source localization.

Underwater, because sound moves 4.5 times faster than in air, it reaches each ear too close together for our brains to discriminate the time between ears. Most fish, being even smaller than humans, would have an even harder time. Yet fish had to be able to tell direction somehow. The whole point of hearing was to sense the world from a distance. Hearing would be useless without the ability to pinpoint the source of a sound. And experiments to date showed that fish clearly could tell the direction of sounds.

In 1975, Popper was using a scanning electron microscope powerful enough to 3-D-image the hair cells of a lake whitefish, and he saw some of the hair cells were oriented differently. Recall that fish were the first to evolve ears, and sophisticated vertebrate hair cells, whose ciliary bundles bent along an axis.

In the lake whitefish, Popper saw, the hair cells seemed to be "facing" several different axes. What if the hair cells that were oriented in one direction fired one set of nerves, or a particular kind of impulse, and those oriented on a different axis fired another? This could mean the fish's brain could tell direction, depending on which cells fired.

But testing this hypothesis would not be easy. It meant measuring the fish's nerve impulses while playing sounds from precise angles around its head. In the 1980s, Fay devised a clever apparatus to do so, building on similar devices pioneered in Europe.

Fay reasoned that if he could shake the fish relative to the water in three directions—up and down, side to side, he could stimulate each "set" of directional hair cells in isolation. And luckily, Fay's brother was an aircraft machinist, with access to precise machining tools.

Fay created a table with a small Frisbee-sized pool on it and tracked down industrial vibrators that could create minute, nanoscale vibrations. He placed these around the pool, on the quarters, like 12, 3, 6, and 9 on a clock face. He set the table on a shaker that could move up and down. Vibration in three dimensions, achieved.

Fay then put a goldfish in the center of the pool and measured the nerve impulses from its ears as he shook the fish minutely, vibrating it and thus simulating a sound moving through its body on one, then two, then three axes. Sure enough, different impulses fired with different directions. The fishes' ears detected the particle motion of sound, and the hair cells sent different responses when the hairs bent in different directions, which its brain could then discriminate, allowing it to tell direction underwater.

IN 1987 SCIENTISTS Bernd Ulrich Budelmann and Roger Hanlon wrote a paper charmingly entitled "Why Cephalopods Are Probably Not 'Deaf'" and pioneering studies were done by a handful of researchers including Sven Dijkgraaf and others in Europe, yet still there was relatively little work on invertebrate "hearing" until the 1990s, when the SOSUS listening network emerged from secrecy.

SOSUS had tracked submarines around the world for two decades, listening to the sounds made by their propellers. But unknown to the United States, there was a spy ring in their midst. Two naval officers, John Walker and Jerry Whitworth, for years had been passing information about naval capabilities, including SOSUS, to the Soviets. In response, the Soviet Union engineered ever-quieter subs that were much harder to detect by passive listening. In 1985, the spy ring and information breach was discovered and SOSUS's secrecy was compromised. The Cold War ended in 1991, and the hydrophones' existence was declassified.

Reports from as far back as World War II were finally opened, and many civilian biologists were delighted to discover that this network had yielded the first recordings of humpback songs, fin whale calls, and fish choruses. The influence of SOSUS on marine bioacoustics is difficult to overstate during its early years as well as when it became broader knowledge. For it was after declassification that it inspired civilian scientists to listen closer in their own work.

In 2022, Simpson's team wanted to confirm just how baby corals were detecting reef sound. They looked (much) closer at the cilia on the surface of free-swimming baby coral, using high-powered microscopy. They had some preliminary hints.

"We can see the hair cells on the outside of the surface of the coral larvae," Simpson says. "When sound passes through the coral larvae, the hair cells start moving differently. So, they're tuned in to specific frequencies of sound."

It's important to note that these cilia are not the same as the hair cells found in vertebrates' ears, or invertebrates' statocysts and other organs. But they are still mechanosensory cells, and able to respond to the motion of water.

Hearing may have first evolved to sense information, but evolution is not wasteful, and many animals evolved the ability to use sound for a fundamental function of life: communication. Along with reefs and waves, animals listen to each other.

Communication can become very complex. But at its core, it's often basic information like *who*, and *where*, packaged in signals designed to get the message where it needs to go. Aquatic animals communicate with sight or scent, touch or taste: Yet just as with hearing, animals' sounds also gain relative importance beneath the waves.

Some of the masters of underwater communication are those with the simplest "voices": fish. They make sound in an astounding variety of ways and are very clever about it. Though whale sounds may be the most famous, fish sounds evolved much earlier, and their sounds demonstrate communication principles perhaps clearest. Among the first scientists to really listen to them was a name we've come across before.

CHAPTER 4

Conversations with Fish:
Communicating in a World of Sound

In the fish world many things are told by sound waves.

—Rachel Carson

BACK ON THE EVE OF World War II, a group of businessman-adventurers and film buffs (including Leo Tolstoy's grandson Ilya) decided that they wanted better movies. Specifically, they thought there should be somewhere to film good underwater footage of charismatic whales and dolphins. So, in 1938, the group funded a movie studio-cum-theme-park in St. Augustine, Florida, called Marine Studios. It came complete with motels, a Greyhound stop, and the Moby-Dick Lounge, where Hemingway was known to drink on occasion. The filming site for portions of *Creature from the Black Lagoon*, it was the first commercial "dolphinarium" and hosted dolphin shows for the public.

Scientists also studied other sea creatures at Marine Studios. In 1952 biologist William Tavolga had come to Marine Studios with his wife, Margaret, a biologist and professor who was studying social and maternal behavior in bottlenose dolphins. Tavolga himself was studying the frillfin goby, a local resident of Floridian tide pools. Occasionally he would show lab visitors what happened when he dropped a female goby in the tank with a male. If the female was "gravid"—swollen with eggs and ready to mate—the male would put on quite a display. His body would turn a lighter color while his chin and neck flushed black as he shook his head rhythmically.

One day, Tavolga showed this display to his colleague Ted Baylor of WHOI, a "great gadgeteer" and audiophile who happened to be carrying a microphone, recorder, and amplifier. After seeing the display, Baylor asked Tavolga if the goby was making sounds.

"But fish don't make sounds, and they can't hear," Tavolga said, echoing the general state of fish-sound knowledge at the time. Baylor insisted they listen, and when the pair wondered how to waterproof his microphone, its size and shape, Tavolga writes, "was phallic enough to suggest an obvious solution: a condom."

The duo heard the male making little grunts in perfect timing with his head shakes. Tavolga had never considered sound a part of fish biology, but now it piqued his curiosity. How did fish sounds work in context, in their lives? In short, how did fish use sound to *communicate*?

SOUND IS A great way to communicate underwater. Sound can be modulated by frequency, duration, timing, and amplitude. Depending on the animal's auditory capabilities it's possible to discriminate one tone or sound from others. Humans may be primarily visual creatures (though this is subjective, and probably cultural), yet we too use sound for a key human communication trait: language.

Below the waves, it was becoming known that many more animals than just marine mammals communicate with sound. But Tavolga's experiments with gobies were among the first to ask questions about what that important ability truly *meant* to fish. His experiments were the first to try to understand their biology, their motivations, and their critical mating rituals by listening to their sounds.

Off a coastal causeway near Marine Studios, Tavolga gathered gobies from the tide pools during summer spawning season. He put twenty males each in their own tank and leaned a flooring tile on each wall for a nest: Many fish parents don't care for their offspring, but in species that do it's often the male who makes a nest and tends the eggs. Tavolga knew that during mating season, when male gobies saw another fish, they either tried to fight them or court them. What senses were they using to distinguish a rival from a potential mate? And how did sound come into it?

Communication, in its most basic definition, is a signal from a sender to a receiver, where both get some benefit. Among the questions these signals answer, two of the most common are: *who* the sender is, and *where* the sender is. Sound is especially useful for alarm calls because it works over distance and around obstacles. *Who* and *where* are also critical in finding a mate.

And reproduction is one of the primary functions of life. Fish must send signals of some kind that another fish can sense. Something that communicates the who, and the where, and that might draw another fish's attention.

Tavolga figured that the fish could be using three senses to communicate: sight, smell, and hearing. He wanted to untangle all three. First, Tavolga took sound and chemicals out of the equation. He lowered another goby, enclosed in a jar, into each test male's tank. Sometimes it was a gravid female; sometimes it was a non-gravid female or another male. The only sense with which the resident goby could identify the stranger was sight. The males' reaction was divided. Some males tried courtship. Some tried combat. Some tried both. They never responded to dead or anesthetized fish, or to different species. This told Tavolga that the sense of sight cued the males to respond, but not exactly *how*.

Tavolga then tested the role of scent, or chemicals, in the water. He took pieces of tissue paper soaked in various substances from other fish and exposed them to the resident males. Blood, urine, and feces got no response. Tissues swabbed from the skin, anal, and genital areas of females also got little response, with one exception: tissues from gravid females' ovaries. Males started flushing and nodding. To confirm it was a chemical response, Tavolga plugged the fishes' nostrils (yes, fish have nostrils) and the response stopped. This showed him that smell triggered the males' display, but not at the right target; they were displaying to a tissue. What about the sounds?

Tavolga turned to the females. They weren't giving the male many obvious cues. They didn't seem to grunt. Their color never changed. But they were reacting to his cues: When they were played recordings of the males' grunts, they moved toward the sound source. Once a female was close enough for him to see her, a male's head would nod and his color would change, possibly to show her where to focus.

The fish were using each sense at a different stage of the dance, at a different range, depending on how that sense best worked underwater. He smelled her and saw her and started grunting. She saw him and moved toward his sound and color change. They then coordinated grunts and nods until they spawned.

This complex ballet had a purpose—coordination. Fish don't have penetrative sex: They release their gametes side by side. And if you've ever spilled something in water, especially moving water like a river or an ocean, you know that if you release two things and want them to have any chance of mixing, you need to release them simultaneously.

Mating involves some truly impressive sounds in fish. Perhaps the best-studied such sounds are made by a fish that happens to frequent the coasts of my home, the Salish Sea and Puget Sound.

I first heard about the plainfin midshipman from Kieran Cox years ago. Back then, he was a PhD student, helping with the intriguingly titled BC Fish Sound Project with lab co-mate and acoustician Xavier Mouy, developing underwater hydrophone and audio arrays. If you listen at the right place and time, in springtime around Vancouver Island, Cox told me, you can hear the midshipmen. Since that conversation, I've wanted to meet them. Luckily, a leading midshipman researcher is just down the shore from me.

"I WAS TALKING to Art Popper yesterday," Joe Sisneros says conversationally as we wait for the elevator. Sisneros's deep voice carries without apparent effort, the kind of voice that could get everyone in the party to move out onto the deck. Even standing before him, I still don't know exactly what he looks like, because he sports the required Covid mask (gray, coordinated nicely with a purple button-down and gray slacks), so I am paying outsized attention to his voice. Masks have re-weighted my perception of social cues, and without being able to see as much of people's faces, I'm listening to voices more closely. I wonder if the pandemic has made the world a bit more oceanic.

Sisneros is a biologist at the University of Washington in Seattle, and he studies the plainfin midshipman, a West Coast fish with a booming voice. Like Tavolga's gobies, or the toadfish whose humming in the Chesapeake had so alarmed the hydrophone operators in World War II, the midshipman hums to find a mate.

In a room down the hall from the elevator, burbling tanks are stacked. I peer closer at one, where a midshipman hovers, docile on the gravel tank bottom. I feel like I'm meeting a celebrity. And I feel somewhat chastened: It has a distinctly unimpressed look.

That may be because they hum at night and it's currently day, though a dim, rainy, November one. Or it may be that their wide downturned mouth and bulbous eyes have a perpetual look of accusation, a fishy case of resting bitch face that straddles the line between cute and ugly. In lieu of scales, the little fellow in the tank has mucus-slick skin the color of coastal muck and a large triangular head melding into a diminutive, tapering body. Petal-shaped fins riffle the sand. If the fish were to turn over, neat rows of photophores—light-emitting spots—outline its belly and abdomen, evoking the gleaming buttons on a Royal Navy midshipman's coat. But it is not its little buttons I came to be impressed by: it is the voice.

"The males during the breeding season produce this call," says Sisneros. "It's an advertisement call." They spend their year out in the sea, hovering midwater. But as spring advances, the males go ashore, searching out muddy flats studded with stone. They wiggle out a nest beneath rocks, *above* the low tide line. They settle beneath the rocks, where in pools of retained water, they'll absorb oxygen through their slick skin.

In the male's large body, the muscles bracketing his swim bladder have swollen immensely. Sisneros shows me a picture of the muscles adjacent to a female's, and then a male's, swim bladder during the spring breeding season. It's like comparing the thighs of an average human and those of an Olympic speed skater. "They bulk up like a bodybuilder," Sisneros says.

As darkness falls the need to call overwhelms the risk of being exposed on the shore or the shallow water. Out in the water—somewhere— are females, ripe with eggs, and they will never find him if they cannot hear him. Finally the male's muscles vibrate and he starts to hum.

"This male is contracting his muscles one hundred times a second," Sisneros says. "And they can sustain that for an hour, two hours. We don't know of any other animal that can contract their muscles at that rate for that length of time." Midshipmen also have swim bladder extensions toward their ears that enhance their hearing, but so that they don't deafen themselves, the twin extensions stop just short of the inner ear.

The hum is a show of stamina. A bigger fish with bigger muscles and a bigger swim bladder vibrates louder and the sound carries farther. Midshipman males, like gobies, will guard the nest until the eggs hatch, and won't leave even to eat. A female's investment in the next generation are her big nutrient-rich eggs. The male's investment is his physical strength and endurance as he guards the eggs, and he must signal this acoustically, simply and over distance. So he hums as hard, as loud, and as long as he can.

The sound is impressive. Appropriately, on summer nights in the mid-'80s, right around when "Sausalito Summer Nights" came out, houseboat owners in Sausalito just north of San Francisco complained about a deafening seasonal hum in the water. No one knew what it was, so theories abounded. (The wartime toadfish in the Chesapeake were, for various reasons, not common public knowledge.) The hum's loud, droning buzz spurred guesses from aliens to industry to secret military experiments. Sometimes it was so deafening no one could sleep. Finally, John McCosker, director of the Steinhart Aquarium, speculated it was fish. He took out a hydrophone and dropped it in the water and recorded that harmonic hum, confirming the midshipman as the culprit.

IT'S EASY TO understand why that hum might be mistaken for aliens or military applications. It's akin to a low-flying prop plane. Though the sound is impressive, I don't, personally, find the monotonous drone particularly romantic. But then I am not a female plainfin midshipman, who are very attracted to the sound.

"The females, the idea is that they innately respond," Sisneros tells me.

Sisneros guides me to a comfortably cluttered lab replete with black benches, concrete floors, and Elvis posters on the wall. When he shuts the door, glossy 4 × 6 photographs are pinned to the back. He points out Arthur Popper and Richard Fay.

In the early 2000s, Sisneros, Fay, and others traveled to California,

to a known gathering spot for midshipmen. They collected some females during the breeding season, put them in a large choice chamber
with a high-quality speaker, and played a male's mating call to see where
the females swam.

Most zeroed in on the male's hum, no matter the direction.

This is remarkable, for when the males are calling, they are either in
a puddle or rather shallow water, and their call is fairly low-frequency,
just a few hundred hertz. Sisneros whips out his phone and does some
quick calculations: The male's 100 Hz hum has a giant wavelength of
about 15 meters in water.

In underwater acoustics, there's something called a cutoff frequency.
That's a frequency of sound too low, and thus a wavelength too long, to
propagate simply in shallow water. Fifteen meters just doesn't *fit*.

"This is what got me really interested in this," Sisneros says. "This is
the environment in which they spawn, right?" His finger taps a photo of
the rocky beach. Then he shows me a graph of the midshipman's hum
at different frequencies. I expect to see one peak at 100 Hz. Instead, the
line is rhythmically spiked, like a heartbeat.

The first peak is the fundamental frequency at 100 Hz. But the line
on the graph peaks again and again, a chain of other frequencies. As he
drums, the male's sound generates *harmonics* at multiples of the original
frequency. If the original sound is 100 Hz harmonics can form at 200,
400, and so on, folding higher frequencies into the same sound

"Look," Sisneros says with a glint in his eye, pointing to the 100 Hz
peak. "This is going to get cut off, and what's going to propagate are the
harmonics." Okay, so the male is making a sound whose harmonics will
spread beyond the beach into the deep water where the females wait.
But that's not all.

Sisneros shows me another graph. This one has several staggered
lines on it. Each line is an audiogram of a female midshipman, taken
at a different point throughout the year. For most of the year, her hear-

ing curve matches the male's fundamental frequency—best at about 100 Hz, ideal for conversational grunts in deep water. But as spring advances, the female's most sensitive hearing skews into higher frequencies.

Sisneros explains that as the female's sex hormones spike, they actually change the sensitivity of her ears, making her more sensitive to certain frequencies. As the male midshipman bulks up, the female undergoes an *auditory* shift. She sensitizes to his harmonics.

Before I leave, Sisneros takes me to the lab's back corner. In a glass-walled chamber sits Richard Fay's original shaker table, the same one he used to demonstrate the directional hair cells and their associated nerve impulses in the goldfish ear. The little pool in the middle is the size of a springform cake pan. The shakers are jelly-roll-sized black cylinders, at each quarter.

"We just always kept it up," Sisneros says, gesturing to a block of old computers behind the table. These are the table controls, and they still run DOS. The lab turns it on every year. He could boot it up, make the table shake, but I'd see nothing. Its movements are on the scale of sound, invisible to the naked eye.

I WANT TO hear midshipmen hum in situ. In preparation for spring, I buy a film-cannister-sized hydrophone on a 10-meter cord, and a recorder with headphones. I ask Kieran Cox about locations. He points me to a paper by his colleague, acoustician William Halliday, who reported the dates and times of plainfin midshipman choruses around Victoria, BC.

One chilly spring sunset, I crouch on a pebbly beach above a meter of crystal-clear water, listening. I detect some clicks, like those of knitting needles, and little scrapes. Two dime-sized crabs face off on a pebble, their mottled gray-and-black bodies blending into the sand as they raise their pincers. They circle like boxers, striking each other again and again. With just sight, this interaction would be a mute tableau. With sound, the crabs' drama becomes gripping, real.

As midshipman season approaches I head to Halliday's first listed location, a public dock in a neighborhood just north of Victoria. Dropping scientific hydrophones into the water near a recreational sailboat mooring could be seen as creepy at best, I realize. Above me, a resort's balcony spills down light, music, and jubilant voices. Mating rituals of a different kind. But in the water, I hear just the clicks and crackles of underwater invertebrates.

Halliday's other coordinates take me an hour north up the coast. I arrive at dusk as kids and parents scatter at a park along the shore. At the waterline I spot a shape familiar from Sisneros's lab: It's a midshipman and I approach with caution. But this one is dead, its body dry and desiccated. When I turn it over, there're its telltale buttons. Standing on the shore I spot another floating body slicking gently back and forth between two rocks. And another. There are dead midshipmen everywhere. It's a bloody reminder that these fish are risking their lives coming into such shallow water and onto the shore, where herring gulls and other birds of prey wheel overhead.

I only hear the hint of a hum, once or twice, before the park gates close. There's at least one big reason why humans rarely listen to the ocean: Coastal access is often restricted, expensive, or troublesome. I'll have to try again later.

RODNEY ROUNTREE DRAGS a wheeled handcart out on the Cotuit Town Dock and unloads his kit: a hydrophone in a pelican case, over-ear headphones, a folding lawn chair, a bucket, a fishing pole. Rountree (whose sweatshirt reads ON A CLEAR NIGHT, I CAN HEAR THE FISH LAUGHING) baits a black mesh crab trap with a small can of squid chum and tethers it to a salt-pale wooden post. Through the hydrophone comes the swish-plunk sound of the trap splashing into Nantucket Sound. It's a June evening on the south shore of Mashpee, Massachusetts, the sky is cotton-candy-colored, and in the waters around us, cusk-eels are mating.

Rountree is known as the Fish Listener. Jovial and gray-haired in a bright Hawaiian shirt, he's an adjunct professor at the University of Victoria and often works with Francis Juanes and his lab. He's based in Cape Cod and has listened to fish for more than three decades. His methods are mostly simple: go into the field, drop a hydrophone, and record. He has spent years listening off boats, coastlines, and docks, in deep seas, in freshwater and salt, and developing hydrophone arrangements, recording tactics, site selection, and "the long-term, tedious, detailed work" of the field naturalist.

"I'm pulling out all the stops to see if we can audition something," Rountree, a native of North Carolina, says in his soft drawl. He unloads a paint bucket, in case the trap yields something. I hear a few muted boops, which Rountree says are oyster toadfish, a relative of the midshipman. But the clearest signals, the ones we've come to hear, are cusk-eels. And I hear them right away. They have a chattering call, staccato bursts, like wind-up dentures.

Cusk-eel are not a single species, but several hundred or so. They're bottom-dwellers that live throughout the sea, from shallow water down to deep ocean trenches. They are eel-shaped but are fish, about 25 centimeters long, and like the midshipman, they make sounds by vibrating muscles against their swim bladder.

As the sun sets and we listen, Rountree tells me this dock is where he first realized that cusk-eel lived in these waters, by inadvertently recording their sounds.

It was the cusk-eel that first piqued Rountree's interest in fish sounds. A graduate student at Rutgers University in New Jersey in the late 1980s, he was working with a colleague who mentioned to him that cusk-eels made sounds. Intrigued, he captured some of the fish from the warm water intake of a coastal nuclear power plant where they sometimes congregated. Together they studied the fish in the lab. Cusk-eels spend the day living in the sediment and, at least in shallow waters, only come out at night. To Rountree's surprise, they heard eels making sounds while buried in the sediment.

He started reading up, beginning with Marie Poland Fish's collected studies, *Western North Atlantic Fishes*. He found recordings of cusk-eels taken near Woods Hole in the 1950s, though the sounds were then labeled erroneously as those of a sea robin. It was the same circular problem that's plagued researchers since the days of the "biologicals": If you don't have a sound ascribed to one fish, and you can't see the source, you can't confirm who's making the sound.

In 2001 Rountree surveyed Stellwagen Bank, just north of Provincetown, with Francis Juanes, and they came up with the idea of triangulating the fish sounds using a small array of several hydrophones and a camera. With sound alone, you couldn't figure out the source, nor with video alone: Fish, unlike humans, don't move their mouths when they "speak." But an array of hydrophones separated by a meter or two and accompanied by a camera could triangulate where a sound originated. By cross-checking audio with video, hopefully, you would see a fish at the "source."

After the Stellwagen study Rountree bought a folding clothesline tree, stripped the lines from the tetrahedral metal frame, lashed a hydrophone to the end of each arm, and added a camera. Later iterations evolved into cubic frames of PVC piping. It was off this very dock, he says, that he tested these prototypes, and it was this design that Kieran Cox was testing back in 2016 when he first told me about fish sounds.

Through our headphones now comes the clinking chain of a nearby buoy, along with some mysterious sounds: A distant bass string plucked. A growl, just loud enough that we both look up, surprised. Rountree isn't sure what it is. Even after all his auditions and arrays, most sounds remain mysteries.

The trap does, eventually, yield a furiously squirming cusk-eel. Rountree gently guides it to the water-filled bucket, drops in the hydrophone, and together we listen. Nothing. He stirs his hand through the water, gently picks up the eel. Nothing. Ah, well. Not every fish wants to be a star.

Back at Rountree's sprawling home, in a verdant cul-de-sac, a small plastic-sheeted greenhouse stands in the driveway. As we pull up, I see it's a white PVC frame from one of the underwater array prototypes, now sheathed and repurposed for seedlings.

Over the next few days, I pace Cape Cod's sandy, sun-drenched shores, hydrophone coiled at the ready. In warm summer rain, halfway down the Provincetown jetty, only invertebrate crackles and clicks. Off the beach at West Dennis, the hiss of sand and waves. Near the Falmouth Yacht Club, just the mosquitolike drone of boats. I rent a paddle board at the Bass River, slide my hydrophone into my backpack, and head seaward.

The water is murky and nearly opaque. I don't realize I am paddling over a sandbar until the board's fin scrapes the bottom; when I step onto a knee-deep sandbar near the river mouth, I can't see my feet. Any fish trying to communicate here would have to use sound.

And that might be tricky. On this bright, warm day with a stiff breeze, motorboats cruise constantly back and forth past kayakers and paddle boarders like me, and I brace against a jostling wake every few seconds.

Yet between each motor, I hear it: the cusk-eel's telltale knock. I drift down the river, hydrophone trailing behind me. In the vessel-clogged waters directly opposite the Bass River Yacht Club, the cusk-eel chorus rises highest, each voice chattering over another, accompanied by a gentle chorus of toadfish boops, layered like synthesizer notes. It's beautiful. I stop, glide, listen. Fish are singing, even here.

Perhaps the noise doesn't bother them. But then, it's mating season in their murky home, and sound is how they find each other. Perhaps they don't really have a choice.

NOW THAT I have listened to fish, I do feel like I know them a bit better. But the realities of scientific funding mean that just because something is wonderful, it's not usually enough to justify studying it. Funds are tight, and research priorities are too.

Cusk-eels, for example, aren't the sort of animals that demand a lot of research attention, beyond the few fishers who notice sound through their hulls. Cusk-eels aren't of great public interest, at least in these waters, Rountree says. "They're not a fisheries species." Nor are they charismatic or cute, like a seahorse or a clownfish, in a way that might endear them to the public. Which leads to an uncomfortable question: Beyond an interesting factoid, who cares if the cusk-eel sings? What value does their sounds provide?

One obvious application is easy monitoring for conservation, environmental assessments, or invasive species. Monitoring a species can be labor intensive, especially an underwater species. But listening for that species can be cheaper and simpler.

ROUNTREE ASSUMED HIS hydrophone would be stolen. It was a brand new top-of-the-line model that recorded in .wav format and not the compressed .mp3, up to 10 gigabytes. And his student had set this expensive device in plain sight, just beneath the waterline on a dilapidated pile on Pier 26 in Tribeca, New York City. The pier was a marine field station for the River Project, which aims to restore the Hudson River ecosystem. In 2003 Rountree and his student Katie Anderson put one hydrophone here, and another hydrophone 153 kilometers upstream at Tivoli Bay, to do an acoustic survey of what fish were in the well-known river. The hydrophones took a recording each evening, from July to September when fish sound was likely to peak, and there were fewer boats zooming past at night. Leaving the case overnight increased theft risk. But the case remained.

When Rountree and Anderson pulled the data, amid the cars, sirens, and city noise they heard a slew of biological sounds even in the middle of New York City. Rountree had assumed the data would be just noise. "You don't have to go to the Galápagos, you don't have to go to the Mid-Atlantic rift. You just go to the docks in New York City. And it's an unknown world." Rountree classified forty-four sound types from the Hudson hydrophone. Only two were known fish species. But only three were human-made. Fifteen other sounds were biological (not fish), and the rest were completely unknown. Fish? Invertebrates? Who knew?

Anderson moved on in her studies but Rountree presented the data at various conferences. In Florida biologist Grant Gilmore approached Rountree and asked about one of the "unknown" sounds Rountree had played during his talk. Gilmore said he thought it sounded like some species of drum.

The average freshwater drum is football-sized though it can grow bigger. It lives in rivers west of the Appalachians down through Central America. And it wasn't supposed to be that far up the Hudson. The drum was a non-native species. Fishermen in the tidal Hudson had found drum in their crab pots and nets, and biologists studying plankton off Long Island had found drum larvae. But not, so far, up in the river's watershed.

How it may have got there, however, was clear: the Erie Canal, built in the 1800s across northern New York state as a transportation link between the Great Lakes and the Hudson River. Rail links had since made the canal obsolete for its original purpose, and it's now used mainly by recreational boats. But where two major watersheds are artificially linked, so are two different aquatic ecosystems, and species can jump borders.

No one knew how far into the Hudson the fish had invaded. Monitoring a species along the Hudson's length meant costly man-hours for field surveying and data processing. Rountree wondered if he could slash time and budgets by just listening to the fishes' sounds.

The species had no digitally available "audition" recording in Fish's archive or elsewhere. So, to confirm the fish was a drum, Rountree went to Nashville, Tennessee, on a September evening in 2003 and paid a sport fisher to boat him out to a known drum spot. As the sun set, the fish started calling. Rountree recorded three hours of drum sounds: same as the Hudson sounds. Confirmed.

Then he started in Lake Champlain with a hydrophone and headphones and worked his way along the Erie Canal down to New York. He listened off piers, boats, and shores, using sound to monitor the spread of an invasive species in one of the country's most important river systems.

Listening to fish can tell us more than their communication or mating rituals. It can reveal species even in unexpected places and can help figure out who is swimming where. That's great for researchers. But there's an industry that sees economic value in tapping into fish communications: fisheries.

ATLANTIC COD ARE among the most valuable fish in the North Atlantic. They can exceed a meter in length and outweigh a toddler. In Scotland, along with their close relative haddock, Atlantic cod support a huge fishing industry. In the 1960s new fishing boats were plying the waters, their new engines roaring, the trawls grinding the bottom, the hulls resonating in the fishes' low-frequency registers.

The Scottish government wanted to know: Were the boat sounds bothering the fish? They approached two young researchers, Colin Chapman and Tony Hawkins, who were already studying fish sounds and asked them to do some experiments to find out.

When I reach him in early 2022 to ask him about this work, Hawkins is taking a break from clearing fallen trees around his home in Aberdeenshire after a storm. He's now retired, and his commanding visage and measured tone seems fitting for someone who was longtime director of Fisheries Research Services at Aberdeen's Marine Laboratory, Scotland's head of fisheries research.

Hawkins graduated from the University of Bristol in the 1950s ten years after World War II ended. His interest in fish sounds began in earnest because one of his lecturers had been a wartime sonar operator and had recordings of what he thought were fish sounds. He asked Hawkins to go out with a hydrophone and investigate.

Hawkins had grown up on the coast at Poole, in Dorset, and he'd sailed since he was young. He took a hydrophone back home to the Dorset seaside, set out, and listened. He heard plenty of sounds from Devon all the way to Cornwall. But he didn't know what species were making the sounds. Hawkins needed to listen to fish one-on-one, and for that he needed an aquarium. Hawkins headed north to the University of Aberdeen where he met Chapman, also studying fish sounds. The duo were the first to record haddock sounds in captivity.

"The laboratory knew that Colin and I were working on fish sounds, and they said, 'Do some work on the *responses* of fish to underwater sounds,'" Hawkins says.

In 1964, at the UK government's request, they studied the sound of a fishing trawler in the sea and did some initial studies of fish responses to noise in large tanks. But it was difficult to simulate natural sounds in tanks. Tank acoustics meant distortions, and particle motion (which fishes perceived) was decoupled from sound pressure, meaning that hydrophones weren't recording the aspect of sound the fish were hearing. "So," Hawkins says, "we decided to do some work in the sea." In the open water, experiments would be much more true to real life.

Scotland's steep glacier-scoured west coast resembles that of Norway or British Columbia, carved by long crevices flooded with deep water. Where these scars form freshwater lakes they are called lochs; fjords which reach the ocean, are sea-lochs. Loch Torridon was one such.

"We chose Loch Torridon because it had this deep water close to the shore," Hawkins says. It had lots of fish and few boats in 1966. Here, the team could place ultrasonic transmitters in the fish's stomach or implant them into the abdomen. These tiny devices let out very high-pitched pings that were picked up by hydrophones on the loch floor. The transmitter pings graphed the fish's positions in response to noise. Did they flee? Surface? Dive? Day or night, Hawkins says, "we were able to track them in three dimensions, up and down and sideways."

In this outdoor laboratory, the duo and their collaborators studied how various species moved in response to the sound of fishing boats and other noise sources like air guns; fish hearing curves; and how noise might mask their calls. Their sounds could reveal where and when they spawn.

Off Scotland the fish come together in the spring in large aggregations to breed. During this time, they're distracted, and, Hawkins suspected, less likely to flee a net or a motor. Fishermen had an interest in knowing where spawning takes place. Some wanted an easy catch, but most wanted to know when to leave the fish alone to ensure the next generation.

During spawning, the males live on the seabed. Each individual male makes sounds. The females are up in midwater. They listen to the sounds, and swim down to the males whose sounds attract them. Then the duo swim up into the water column. At this phase of the hour-long courtship, Hawkins heard the male make a series of increasingly rapid knocks. "Then he embraces her and they stop making sounds," Hawkins explains. "And she releases the eggs and he releases the sperm." They are coordinating their movements to ensure they release their gametes at the same time.

Hawkins's 1967 paper on these knocks made the cover of the prestigious journal *Nature*, accompanied by his line drawing of the spawning haddock: two fish swirling together in the water.

I find these haddock oddly beautiful, with big eyes and leaflike fins. I note Hawkins has image credits for diagrams and illustrations, and he laughs: Yes, he loves drawing. He's illustrated much of his own work and others'.

To show that fish sounds were useful beacons of spawning fish in practice, "We went to a little place in the north of Norway," Hawkins recalls. He and his collaborators took a boat along the coast, listening. Haddock sounds, which Hawkins knew so well by now, indeed pinpointed these aggregations, providing data so the Norwegians could leave the fish alone to ensure next year's catch.

Fish sounds—fish communications—are a window into lives. They can help us understand what a fish wants, and how it's trying to get it. But sounds can also be a valuable monitoring tool for fisheries and even in tracking invasive species. And listening to fish like the humble cusk-eel, not just the superstars, is a useful measure of one of the most critical measures of ecosystem health.

IT'S A DARK, still night in late May, just north of Victoria, BC, the second year of my quest to listen to plainfin midshipmen. My hydrophone cord is coiled in my pocket, my headphones (over-the-ear to block noise, as Rountree suggested) around my neck, as I weave my way along dark paths to a long public pier in a quiet cove, in the heart of this small seaside community. At ten p.m. all I can hear is the distant sound of voices, people on decks or moored in sailboats, and the quiet kiss of water on docks and hulls. The tide is low, and the wooden stilts hold me high above the water. Hoping no one wanders down here to see what I'm doing, I hang the hydrophone over the railing's edge and spool it out, down, down, until in my headphones I hear a little *plink*. It's in.

And there is the hum, clear and unmistakable. The plainfin midshipmen are here.

The grin that breaks out on my face is so wide my cheeks hurt. It's an unremarkable sound but knowing what it *means* thrills me. When I lift the hydrophone from the water the sound instantly vanishes. There's no clue in this calm night air of the drama playing out in the sea as the midshipmen fish sing with all their might, drowning out the little clicks and snaps of the other small denizens going about their lives in the water below.

As far as we know, many invertebrates *are* rather silent. But there are exceptions: The pistol shrimp with its bubble-claw uses sound to defend its territory. Polychaete worms snap their jaws; fiddler crabs snap their powerful claws. Many animals still use incidental sound to communicate with mates or enemies. We are just beginning to understand the true scope of these interactions.

There is another way that animals can use sound, called autocommunication, which literally means communicating with oneself. Animals above and below the waves do this, but here, we turn to marine mammals, some of whom have evolved a spectacular form of this: echolocation.

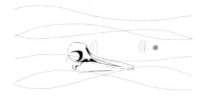

CHAPTER 5

Click to Reveal:
The Evolution of Echolocation

Perhaps it is impossible for us to conceive how
such a brain might perceive the world.

—Sam Ridgway, *The Dolphin Doctor*

LAZARO SPALLANZANI HAD A PROBLEM.

It wasn't the sort of life-or-death problem, like disease or war, that might otherwise confront an eighteenth-century Italian. Spallanzani— the kind of natural historian who studied whether in vitro fertilization was possible, the spontaneous generation of life, and even digestion by swallowing permeable bags of food on strings, and then pulling them back up his esophagus—was vexed by bats.

He first encountered this perplexity of bats in 1793 when he blew out a candle in a room, plunging it into near-total blackness. There happened to be an owl in the room, and despite the owl's reputation for seeing in the dark, the bird careened into a wall. Spallanzani wondered just how good owls' purportedly excellent night vision was. He tested the owl, and several bat species, out of curiosity. When he blew out the candle on a bat, it alone flew on, undeterred. How?

He tried various strategies to solve the mystery. (And let this serve as a warning that his methods were horrifying.) To be sure the bats couldn't see, he blinded them, either burning their corneas with red-hot wire or plucking out their eyeballs and severing the tendons. The bats (once they recovered) still flew just fine in the dark. He wrote to friends around Europe, pondering how a bat could fly blind around furniture and other objects and not crash, as the owl had? The question would become known as Spallanzani's Bat Problem. In one letter he wrote, "There probably occurs to you the thought which first came to my own mind, namely that some other sense takes the place of sight."

Swiss physician Louis Jurine expanded on Spallanzani's experiments. After blinding the bats he blunted their hearing and plugged their ears with wax, turpentine, and other substances. The bats were completely disoriented. Jurine concluded, "The organ of hearing appears to supply that of sight in the discovery of bodies . . ." Spallanzani was electrified that "their ears rather than their eyes serve to direct them in flight."

Debate ensued. In the late 1790s prominent paleontologist Georges Cuvier, whose work comparing fossils across animal groups would inform Darwin's theory of evolution, advanced a competing theory that bats navigated using a kind of touch. Cuvier thought bats' sensitive wings felt subtle shifts of air as they approached objects. He slammed Spallanzani's "cruel" experiments and the idea that bats had some "sixth sense." Spallanzani died in 1799, convinced that bat navigation involved hearing, but he never proved exactly how.

The bat problem went unsolved until the twentieth century. After the *Titanic* disaster, in the rush to invent sea-navigation devices, inventor Sir Hiram Maxim's idea for an iceberg navigation device was inspired by bats' "sixth sense." Maxim thought this sense worked as follows: Bats beat their wings several times a second, and this made sounds too low for humans to hear, but that bats sensed, and they navigated by the echoes. Ships, following this idea, should be equipped with a sound-producing apparatus and microphone. They should send out loud, low-frequency blasts in all directions "constantly" in foggy or hazardous waters and navigate by the echoes.

That Maxim would think up the idea of echoing sound is darkly ironic, because he had invented the first machine gun, the "Maxim gun," in the 1880s, and became completely deaf from the noise of experimenting with it. And while Maxim's invention didn't come to be used, it (and, so, bats) preceded Fessenden's oscillator, which brought human echolocation below the waves. But when scientists tried to record actual bats navigating with sound, they heard nothing.

Biologists Donald Griffin and Robert Galambos finally showed how bats use sound—high-frequency, not low—in 1938. Griffin worked with physicist George Pierce, setting microphones all around a room that recorded clicks above the frequency of human hearing. They heard a stream of these ultrasonic clicks—but only in the bat's direct path. Evolution had beat sonar techs to beams of ultrasound. They coined the word "echolocation," in which animals glean information from the echoes of their own sounds.

When an animal only listens, it's hearing sounds from a source elsewhere. But when it makes its own sounds and then listens to the echoes, through autocommunication, it's actively exploring the world. It's akin to the difference between passively looking around a room in daylight and shining a flashlight into the same room's dark corners and closets. But what can echoes tell us?

The mythological Echo was a nymph who was cursed by the goddess Juno to only repeat the last words of any sentence she heard. She had no words of her own. But a returning echo does bring new information about what it bounced off. A discerning ear can extract this data. As with sonar, an echo's return time tells the distance to the target. And the echo's frequency will shift slightly, depending on whether the target is moving toward or away from you. This is the so-called Doppler effect. If a bat clicks at prey moving toward it the clicks will return at a slightly higher frequency, since each successive click bounces back a bit sooner, squishing the wavelengths together. Conversely, if the prey is moving away, echoes will shift lower in pitch. This effect is also why emergency sirens seem to change pitch as a speeding vehicle passes. (Doppler shifts also happen to light in astrophysics, shifting red or blue depending on how fast a star or galaxy is moving away from or toward us. These shifts helped us discover the expansion rate of the universe.)

Animals who echolocate can read what's around them based on the time it takes the sound to return, and any shifts in pitch help determine whether it's moving. An animal can do this all before it sends out the next click, creating a rhythm of click, read, click, read, click, read. Griffin speculated that in addition to bats, other animals may do it, including marine mammals. After all, they must navigate, eat, and stay alive in deep, dark water.

The answer to the question "Which animals do this and how?" would not be long in coming. For in 1938, that same year, Marine Studios, the first aquarium with dolphin shows, and later the site of William Tavolga's goby experiments, opened its doors.

RONALD V. ("RONNIE") CAPO was a fisherman, a proud Floridian, and really good at catching dolphins. Capo began catching dolphins for the new Marine Studios in 1939, herding them into shallow coves or bays using nets. One day, Capo told the aquarium's curator, Arthur McBride, something funny he'd noted. The nets he used for fishing were fine mesh, with only small gaps between the strands. These nets weren't great for dolphins, which jumped right over the nets' edge and escaped. They did this even when it was completely dark or in murky water.

However, if Capo used a net with bigger holes, some 10 centimeters between the strands, then he snagged the dolphins, though they could have easily leaped over the net. There was only one exception: if one dolphin got tangled and pulled the floats on the net's edge down under the water. Then the other dolphins flocked to the gap created by the lowered edge, leapt over, and escaped.

McBride went out with Capo and saw this for himself in murky waters, and at night, where the dolphins clearly couldn't see a thing. Yet they easily hopped out of the finer net with ease. How did the dolphins see in the dark, murky water? Why did they *not* see the coarser net? And why did it make a difference if the floats were submerged?

"This behavior calls to mind the sonic sending and receiving apparatus which enables the bat to avoid obstacles in the dark," McBride wrote in his journal in 1947. He mused that since sound was clearly important to dolphins, they too might use it to "see" their world.

Yet McBride didn't test this hypothesis. Instead, he mentioned it to others, including William Schevill, who went on to research echolocation himself. But part of the reason people like McBride knew sound was important to dolphins was because aquariums like Marine Studios had begun to familiarize people with dolphin clicks and whistles. This suite of sounds was something that dolphins shared with many other whales. But not all of them.

WHALES FALL INTO one of two fundamentally different groups: baleen whales and toothed whales. Baleen whales—the blue, humpback, fin, bowhead, and right whales—are filter feeders that, instead of teeth, have large vertical plates of baleen, which is made of protein similar to that in human fingernails. Baleen whales take big gulps of water, then sieve the water back out, leaving the prey behind—tiny plankton, fish, and other small ocean denizens. Filter feeding is incredibly efficient and provides a lot of energy, and these whales grew enormously. They make mostly low, loud, sonorous sounds. The toothed whales, the other main group, are called odontocetes. (From the Latin *odont*, or tooth, and *cetus*, or whale.) These are the dolphins, killer whales, beaked whales, pilot whales, narwhals, and belugas, and they are meat-eaters, hunting fish and other marine mammals.

Toothed whales make several types of sound. Some species make distinct sounds in each type. Others, such as belugas, make "graded" or combined sounds, where these types melt into each other. But generally, the sounds fall into three categories. *Whistles* are pure tones, with a very narrow range of frequencies. Sometimes, like the midshipman hum, whistles can have harmonics too. Then there are *pulsed calls* (or "burst pulses" in some species). These are trains of sound pulses that merge in our ears to sound like whoops, groans, and all manner of things. Finally, all odontocetes make *clicks*. Exactly what the clicks were for was still uncertain in the 1950s. But evidence was growing it was for echolocation.

The prospect of dolphin echolocation was particularly enticing to the U.S. Navy, which remained interested in underwater sound. Following their beluga recordings in the St. Lawrence, William Schevill and Barbara Lawrence got naval funding to study the phenomenon. In the early 1950s the team brought a dolphin to a seaside pool near Woods Hole and studied how it detected objects and navigated mazes using both sight and sound. Simultaneously, the Navy-supported psychologist Winthrop Kellogg at Marine Studios in Florida showed that dolphins heard sounds well above the range of human hearing. By 1956 Schevill and Lawrence watched a captive dolphin named Tvas locate and capture fish without sight, and they thought Tvas was likely echolocating.

In 1961, biologist Kenneth Norris, working at the newly opened

Marineland of the Pacific in Los Angeles, published confirmation that dolphins echolocated with ultrasonic clicks by "blinding" a dolphin named Kathy. Thankfully, instead of Spallanzani's barbaric tactics or Frisch's blinding of Xaverl the catfish, Norris simply affixed rubber suction cups over Kathy's eyes. Kathy swam blind through an obstacle course, a pool staggered with vertical pipes, with no trouble, clicking the whole time.

A trainable marine animal with its own excellent sonar excited the military brass at the height of the Cold War, including William McLean, of the Naval Ordnance Test Station at China Lake in the California desert. McLean had invented the heat-seeking air-to-air "Sidewinder" missile, inspired by the eponymous desert snake that used infrared heat sensing to find its prey. McLean knew that technology could learn from animal abilities.

McLean had heard of dolphin researcher John Lilly who, in his book, *Man and Dolphin*, stated that within decades man would communicate with dolphins. McLean wanted to know if this was true, if dolphins might inform better torpedo design, and how their sonar worked. So, the navy opened a marine bioscience facility at Point Mugu in California.

But moving dolphins and keeping them healthy in captivity proved tricky. Even though some aquariums like Marine Studios had kept bottlenose dolphins for decades, there was still much to learn. The naval personnel discovered that because they are adapted for cold water, dolphins lack sweat glands so they tend to overheat during transportation in the air. And if they are not kept constantly wet, their skin rapidly cracks and peels. Dolphin skeletons can't support their own weight out of the water, so moving them also risks broken bones. They learned the animals needed to be fully supported, and kept constantly wet, during transport.

Once inside the tanks, dolphins were often heartbreakingly distressed. In his 1973 scientific memoir, *Marine Mammals and Man: The*

Navy's Porpoises and Sea Lions,, Forrest G. Wood described the first attempt to keep a Dall's porpoise at a naval base in the mid-1960s:

> When released in our 50-foot tank the porpoise swam directly into the wall. It then repeatedly attempted to sound through the floor of the tank. After about half an hour of this kind of behavior it was caught and put in a harness to which a leash was attached. An attendant could then usually deflect the porpoise before it struck the side of the tank. Finally, the leash was attached to a cord stretched across the tank. [. . .] The animal had already severely injured itself, however, and died sometime in the early morning hours. Sam Ridgway, our veterinarian, conducted a postmortem examination and concluded that death was due to trauma and hemorrhage caused by the beating the porpoise had sustained in trying to swim through the wall and floor of the tank.

Ridgway had graduated from Texas A&M as a veterinarian in 1960 and went into the air force at Oxnard Air Force Base. In 1962, he was brought to the Point Mugu naval base, near Oxnard, California, to perform an autopsy on another dolphin that had died in captivity. The role stuck, and Ridgway became the first marine mammal veterinarian, for the first dolphins seriously studied for their echolocation abilities. His work with one dolphin, Tuf Guy (or Tuffy, as Ridgway called him), was the subject of his memoir, *The Dolphin Doctor.* Tuffy and other dolphins were trained to retrieve explosives or other equipment that were lost from vessels, and deliver packages to divers dozens of meters down. (The navy uses dolphins even today for similar "helper" tasks, assisting navy divers with detecting mines and enemy divers. Dolphins have even served during conflicts, including the Gulf War. As for transport, they now travel in fleece cradles suspended in wet boxes.)

The U.S. Navy in 1968 opened the Naval Ocean Systems Center field station at Kaneohe Bay in Hawaii. Studying dolphins in tanks just wouldn't give the same results as studying animals in the wild. Captive animals are expensive and difficult to keep, and there's no way to know if they're acting or eating or making sounds the same as they would in the wild. But by the 1960s, field equipment like electronic satellite tagging improved, and it became possible to follow wild animals, and work in both tanks and in the wild slowly revealed intricate details of how dolphins echolocate.

WE KNOW APPROXIMATELY when echolocation evolved, at least in marine mammals. About 55 million years ago, during the Eocene epoch, the temperature was warmer than today by at least a few degrees, and the average sea level about 150 meters higher. (For comparison, the Great Pyramid of Giza is just 137 meters high: All of Florida, which barely tops 100 meters, would be underwater.) One early ancestor of modern whales and dolphins was a one- to two-meter-long mammal relative of modern hippos that strode on land. Perhaps seeking new prey or a safer habitat, it gradually waded deeper and deeper into the swampy shores of the Tethys Sea in what is now Pakistan, hence its moniker: *Pakicetus*. (Greek: *Paki* = "Pakistan," *cetus* = "whale.")

Ambulocetus (or "walking whale") lived about 49 million years ago. Both animals were amphibious, like modern seals and sea lions. But *Ambulocetus*'s hind legs were beginning to widen into paddles. *Remingtonocetus* still had four weight-bearing limbs, which meant it could still walk on land. But its ear bones had begun to isolate from its skull, suggesting it had already adapted underwater hearing.

Around 35 million years ago lived the *Basilosauridae*, the earliest completely aquatic whales. From this common ancestor, today's whales and dolphins slowly evolved. As whales grew ever more aquatic, along with losing hind legs, their nostrils slid farther and farther up their faces, making it easier to breathe without coming out of the water, and eventually became blowholes.

Around 35 million years ago, whales split into two main groups, apparently divided by something very fundamental: their diet. One group started filter feeding and became the mysticetes: the blue, fin, and humpback whales, the largest animals on Earth.

The other branch kept their teeth and stayed carnivorous, hunting fish and other marine animals. Dolphins, belugas, and killer whales, porpoises and sperm whales, beaked and pilot whales evolved their own feeding superpower: echolocation. But we don't know exactly when.

What we do have are fossil clues from ears. The length of the cochlea, its number of spiral turns, may relate to a mammal's hearing range. Luckily, cochleae fossilize—at least, the small curl inside them do. If you look at the number of turns of the fossilized laminae in a whale ear, you can infer if the animal has evolved to hear high frequencies—that *might* mean it echolocates—though this isn't conclusive. But by around 25 or 30 million years ago, the toothed-whale group could almost certainly echolocate.

Astounding as this ability is, creatures from shrews and mice, and even swiftlet birds, can do so. Even humans can learn to echolocate. Some people who have lost their sight have learned to click and read from the echoes of the objects around them, navigating through rooms and in some cases even biking through cities.

I WANTED TO know more about this marvel of acoustic evolution, so I asked Nicholas Pyenson, curator of fossil marine mammals at the Smithsonian Institute's National Museum of Natural History in Washington, D.C., and author of *Spying on Whales*. He's written about what bones tell us about the evolution of whales' acoustic abilities. In his rapid, enthusiastic speech he drops beautiful and utterly mystifying Latin- and Greek-derived terms with the ease of familiarity, describing bones I had never guessed existed.

Reconstructing an ability like echolocation with bones is a challenge. But the skulls, jaws, and ear bones of whales all fossilize. Pyenson has also worked with researchers who have pioneered detailed computerized tomography (CT) scans of fossils. Although at this moment travel is off the table thanks to the coronavirus, Pyenson can still show me the 3-D virtual skull of a bottlenose dolphin.

With me on my living room couch in lockdown, and Pyenson in a crisp shirt and trim goatee in his D.C. office, he rotates the skull on my screen to show me the two sockets on the side of the skull, where its eye will sit. Then he deftly flips the skull to show me the vaguely ball-shaped indentations on the skull's underside where its ears reside. He points out the skull's long rostrum, the bone in its bottle-like nose, which juts forward like a flat blade. But the oddest feature, I think, is its forehead. Where a human skull would dome outward, the dolphin's bows *inward,* a concave surface like a bowl set in its face. This, Pyenson says, is where the melon sits in a living animal—the melon is the part that bulges like a forehead on a dolphin or a beluga. It's made of soft fat and so doesn't generally fossilize.

Behind the melon on that bony bowl run the nasal passages, connecting its lungs to the blowhole on the top of its head. Along these nasal passages are blind sacs (nasal diverticula, if you want to be technical about it, as Pyenson is). We can't see this in a skull, but in a living creature, these sac entrances are sealed by structures called *museau du singe*—literally, "monkey lips." (This time, in French, not Latin or Greek!) "They will vibrate," Pyenson says, the way ours would against a trumpet. (More recent work suggests clicks are made by the right monkey lips, and whistles by the left in dolphins and false killer whales.)

Now that I'm oriented in a dolphin's head, I can sort of see how echolocation works. "Echolocation in an odontocete is three steps," Pyenson says. First, the click, then hearing the echo, and finally there's processing it in the brain.

Dolphins click near their nasal diverticula with the charmingly named museau du singe. A click is "broadband," or a sound with frequencies across a wide range. (Contrast that with a pure tone of just one frequency, like a piano note or a whistle.) A dolphin's click contains frequencies from the upper edge of human hearing—about 20 kHz—all the way up to above 200 kHz. Most of the click's energy is often in the frequencies around 100 or 120 kHz. This means that while we can hear the click's lower frequencies, we can't hear its higher frequencies. And yet those higher frequencies are necessary for a good click.

As Langevin's "submarine problem" showed, a meaningful reflection off something needs a wavelength about that object's size or smaller. A 120 kHz wavelength in water is about 1.25 centimeters—great for bouncing off of a delicious fish or its air-filled swim bladder. That, and beam formation, is why animals that echolocate need to go high. A click is very short, from several hundredths to just thousandths of a second. Dolphins' click rates vary but they can reach up to six hundred per second and maybe more.

Toothed whales focus their clicks into a beam of sound. Cupping the lips are the concave bowl of the skull, dense connective tissue, and air sacs, all of which, thanks to their impedance mismatch, will reflect the sound forward into the melon just as the cuplike reflector on a flashlight sends light in one direction.

But it's not a beam yet: It passes through the melon, "which is made up of extremely toxic fatty acid chains," Pyenson says. How whales hoard these fats in their body isn't clear, but the fats are never used for energy, even if the animal is starved. Different species have different-shaped melons, but the general principle takes advantage of our old friend, acoustic impedance. Different fats have different densities, from waxy solids to oily liquids, and melons layer these fats. The side and outer layers are a mesh of dense tissue, while the central inner layers are successively thinner, and the center is liquid. The melon curves the sound and focuses it into its beam. The click sounds leave the head, Pyenson holds up a hand to his own forehead, "like a searchlight."

Then comes the hearing. Pyenson fetches a real skull from his office behind him, because the CT scan we're examining doesn't have its minuscule ear bones. Most whale ears are loosely attached and are among the first things to fall out of the skull when a whale dies and the fragile connecting tissue rots.

Pyenson returns with a taupe dolphin skull in one hand and a small box in the other. He lifts a tiny bone, the size of an earbud and shaped a bit like two small bubbles pushed together. He gently holds it in one of two concavities underneath the skull. This small bone is the periotic, he says. He plucks a second small bone. "The tympanic," he says.

The periotic and tympanic fit together like a little puzzle, sitting inside the skull, but not attached. In a living dolphin, they'd be buoyed in a foamy structure that isolates the ears acoustically from the skull. This helps the whale detect sound distinctly in each ear, with no bone conduction. This ear isolation helps marine mammals pinpoint the source of sounds underwater.

These little bones, Pyenson says, are very dense—some of the densest tissue of any animal. Sound travels through them into the whale's middle ear with its three-bone chain—malleus, incus, stapes—and then on to the inner ear and cochlea.

Once the sound reaches the middle ear, it's moving in bone, skirting the impedance mismatch between air and water that plagues human ears. But how does sound get from the water *to* the tympanic?

Odontocetes don't have an outer ear and eardrum. Instead, they hear with their lower jaw. Pyenson fetches a third bone, the flat jawbone, narrow near the chin and wider near the cheek, a row of peg-like teeth embedded in one edge. He holds it up to show the inner side, where it's widest, near the skull. There's a pocket-like opening, like an envelope. The jawbone is mostly hollow. In a living dolphin, Pyenson says, the bone is full of fat. The exposed pad of fat at the wide end connects through the thin area of the jawbone to the hard little tympanic. Sound travels through the water, into the jaw, through the fat, up through the thin, wide jawbone and from there into the tympanic and ultimately the cochlea. No air involved.

The third and final step of echolocation, Pyenson says, is the nerve signal from the cochlea being processed in the brain. Echolocation is not just making a click but the entire process: click, detecting, and processing. And this final step can become truly mind-bending.

SCIENTISTS DIGGING INTO echolocation in the postwar decades wondered: Did dolphin hearing sensitivity shift seasonally? Did some sounds have more meaning, or motivate the animals, more than others? But it would take a lot to get answers to those questions. Wood, Ridgway, and others described the challenges of labor-intensive behavioral studies to figure out what dolphins actually heard, which required quiet tanks, intensive training, and caring for large and often uncooperative animals.

The first dolphin audiograms showed that they could hear from 100 Hz, the low end of human hearing, up to 150 kHz—the widest hearing range of any mammal. They heard best between 40 and 100 kHz, far above human range, and they clicked at a peak frequency of about 120 kHz. This made sense, since they also vocalized and made other whistles and pulsed calls at lower frequencies, and so they could hear each call type distinctly with less chance of masking.

Toothed whales like dolphins have great control over their clicks. One beluga whale that had been swimming in San Diego Bay had made clicks centered on 40–60 kHz. When it was moved to Hawaii, it went up an octave and began clicking at 100–120 kHz, suggesting toothed whales may click at different frequencies in different places. Why they do this isn't clear. (This frequency shift also changes the animals' "beam" shape because as the frequency of the clicks rises, the beam narrows.) One dolphin at Point Mugu, named Scylla (after the devouring sea creature from Homer's *Odyssey*), was trained to echolocate on command and could also change the shape of her click beam.

Researchers found dolphins are very good at distinguishing objects of different sizes. One Point Mugu dolphin, Doris, could tell two objects apart that differed in size by only fractions of a centimeter, as well as objects of different thicknesses, things buried in sand, and objects of different materials: aluminum, copper, brass, glass, plastic. This has to do with the physics of ultrasounds.

A dolphin's clicks centered at about 120 kHz. A modern abdominal ultrasound clicks even higher, at about 2.5 MHz. (If you're wondering why an ultrasound wand gets rubbed against your body, it's because of impedance: no air to reflect or weaken the sound wave.) Medical ultrasound frequencies vary depending on what organ you're imaging. Lower, blurrier frequencies go deeper for abdominal scans, while higher, sharper frequencies reveal things closer to the surface like veins, thyroid, or breasts. That's because lower-frequency sound loses less energy to a medium and so attenuates slower and penetrates deeper. A dolphin uses frequencies high enough to pick out detail, and low enough to penetrate, which means clicking dolphins almost certainly "see" not just the surface of things, but *through* them, reading thickness, material, and other properties.

Is echolocation, then, like sight to a dolphin or beluga? Some research shows that in deaf humans, the brain areas that otherwise process visual information also show activity when they process auditory information. But we may never know what it's like to be a clicking dolphin.

THE CITY OF Anchorage sits near the top of Cook Inlet, a several-hundred-kilometer-long arm of the Pacific nosing northeast into Alaska's Chugach Mountains. It's a working town with industrial bones and stunning beauty, ocean and mountains, iron and concrete woven together. On one of 2018's last warm September days, when the leaves of young aspens in the parks twinkle like coins, I wait for slow-shunting rail cars to move off the road before I drive out to the gravel lot by a small-craft boat launch.

Paul Wade, a marine biologist with NOAA's Alaska Fisheries Science Center, backs a truck carrying a five-meter-long boat with inflatable sides, called a zodiac, on a trailer gingerly down the launch ramp. This is a formidable skill. Cook Inlet's prodigious ten-meter-high tide is falling, and the ramp is lengthening. And lengthening. Wade backs the truck deeper, and deeper. The ramp's lower reaches vanish into pillowy drifts of soft, sucking, dust-colored mud. The tires smear the mud as Wade tries to get the trailer close to the water without miring everything. Beside the dock, deep runnels scar vast exposed mudflats that spread out from the shore.

This ubiquitous mud comes from rivers born in the mountains around the inlet. The mountains give rise to streams heavily laden with sediment. In this inlet, tides up to a dozen meters, the fourth-largest tidal range in the world, drive currents that keep this sediment suspended and turn the water as opaque as a latte as it shifts the muddy bottom like dunes. If thalassaphobia is the fear of deep water, this coast triggers a different fear in me, of mucky bottoms and things at the surface that should remain sunken. It evokes H. P. Lovecraft's *Dagon*, which describes a man drifting in a small boat in the South Pacific and running around on "a slimy expanse of hellish black mire which extended about me in monotonous undulations as far as I could see."

Wade manages to launch the zodiac. He maneuvers it back and forth by the dock in the ten-kilometer-per-hour currents, and we leap over the gap of freezing, muddy sea to get in. The tide is still falling, so we must wait until it turns and rises again for water deep enough to pull the boat back out.

The current is strong, but the wind is calm and there's no chop. The sun's burnished reflection squiggles on the surface as the zodiac streaks southeast, beneath the thundering flight paths of massive planes going in and out of the second-busiest cargo airport in the United States. I spy the ranks of oil rigs in the distance, and the ghost of Denali rising over the horizon. A frigid hour's ride, and we pull up a few hundred meters from the mouth of the Chickaloon River, on the shore south of Turnagain Arm.

Turnagain Arm is a branch at the top of Cook Inlet, and there's a spot near this confluence that Wade calls the "beluga pool." Standing at the boat's helm in mirrored shades and a bright-orange jumpsuit that doubles as insulation, floatation, and flash of color in case of emergency, he slows the zodiac to a polite putt as, around the boat, the surface boils when fifty or sixty beluga whales breach the surface, huff, and vanish.

Wade says these are young animals. Belugas are born gray and only whiten with age, and the tantalizing flashes of gleaming head or flank are indeed the color of storm clouds. Even small glimpses suggest the quick, gawky grace of kids. Sometimes a telltale ripple swirls just off the gunwale, followed by a nudge from beneath and the uncanny feeling of having met someone profoundly new.

The belugas make sound below *and* above the water. Marie Fish, William Schevill, and others recount that sailors of yore dubbed belugas "sea canaries" for their loquaciousness. They may have inspired the mythological sirens of Homer's *Odyssey*. Across the calm water are soft whistles, hoots, and a strange metallic clatter: honks, mewls, and a sound that can only be described as a fart.

The adults swim farther out, like human parents on the sidelines of a splash pad. Their iconic snow-white backs rise, curve, and fall against the water, more slowly than their nimble offspring. Their bodies aren't visible in toto, but we've all seen them: Belugas have lumpen forms with small heads and bulging foreheads, with a smile that gives them that unfortunately goofy visage. Female belugas grow about 12 feet, and adult males can reach 18 feet. Most belugas live in the Arctic but a genetically distinct population swims here on Alaska's south shore.

In the late 1970s 1,300 belugas called Cook Inlet home, yet by 1998 fewer than 400 remained. Native Alaskan tribes stopped subsistence hunting in 2005 despite belugas being an important food. The federal government designated belugas endangered in 2008 but their numbers are still low. Wade is on the NOAA team studying why, including a photo-ID project cataloging who's hanging out with whom, where, and when. Marine mammals are fundamentally social, and understanding them means understanding their social groups.

But the water is most opaque with mud and silt exactly where belugas come to eat in the upper inlet. Here they seek salmon, eulachan, and other fish that swim the Susitna, Beluga, and Kenai rivers. These rivers are skirted with vast shallow mud flats, a risky place to swim. If the water gets too shallow, the air may creep close to a beluga's head or mud too close to its belly. The fickle water levels can easily strand a beluga before it can find deeper water. How could any animal navigate this maze of scouring currents and swinging tides while functionally blind from the murk, let alone hunt in the process?

They are not blind, but if the water clouds their vision, another sense can take its place.

Wade drops a hydrophone connected to a speaker, and more sounds suffuse the late summer air. One sound resembles a kid running a zipper up and down on their coat. Or maybe the starter on a gas stove. These are the clicks, or at least what we can hear of them. But every now and then, the clicks accelerate until they are so close together they shade into a buzz. This buzz is a sign that the beluga is hunting—and researchers can use this sound to study the hunt, even when they can't see it.

MANUEL CASTELLOTE HAS spent fifteen years listening to the Cook Inlet belugas using suction-cup tags and other tools. He's even designed bespoke hydrophone moorings that can withstand this high-energy environment and record whales for months on end. The hydrophones look a bit like mushrooms—he had to give them a low profile so they would neither get stuck in the mud, nor poke so high that they'd be scoured by tumbling ice. In shallower areas where the muddy sediment piles up to 5 meters deep, he plugs them into the mud like thumbtacks.

Using these, he's reconstructed how the belugas use clicks and buzzes when they hunt.

The Cook Inlet population clicks are centered around 100 kHz, which in water has a wavelength of just over a centimeter. This is more than detailed enough to reflect well off of a salmon (60–100 cm long) or eulachan (15–20 cm). Belugas often click in "click trains" from six to more than a thousand clicks. They click slower while they're searching and wait for each click echo to return before making the next. Once they find prey the clicks get faster and become a buzz as the whale starts closing the distance to its target. (The sound varies by species: When Tvas buzzed, Schevill and Lawrence called it "creaks.") The distance between whale and food shrinks quickly, until the sound becomes a "terminal buzz" and the animal either bites . . . or misses.

But belugas buzz to each other socially as well. How can a scientist know if a buzz on a hydrophone is hunting or chatting? In a clever experiment, Castellote inserted a harmless temperature probe into belugas' stomachs that transmitted real-time temperature data to him. At the same time, he recorded the belugas' sounds as they dove. The hunting buzzes ended in a "chomp" but also in a sudden temperature drop in the beluga's stomach from the sudden influx of chilly prey. (When they missed, the sound was an empty "jaw clap.") Castellote split the buzzes into cold, choppy hunting buzzes and social buzzes. He found the latter are less regular, and the click intervals vary more. But the feeding buzzes increase in rate more steadily and reach a shorter inter-click interval. Whales seem to have different clicks and buzzes to suit their needs.

The Cook Inlet belugas in opaque water rarely dive deeper than 100 meters. They can eat in very shallow water, but usually dive for two to four minutes in 3 to 10 meters of water. They don't need to echolocate things that are particularly far away. Dolphins also feed largely at the surface, and while their clicks are intricate, they don't need to carry too far. But odontocetes have also moved into the open ocean—and the very deep. And it was here that humans first encountered clicks in perhaps the most extreme odontocete clicker of all—though they didn't know it at the time.

BEFORE STEAM AND diesel ship engines drowned out the sounds, whalers were among the best sources of natural whale history, including sounds. Whalers spent years at sea, with only the thin hull of a wooden boat between their ears and the water. They heard humpback song, beluga chirps. But the sperm whale had a reputation for silence.

Before the discovery of petroleum, whale oil ran the world. It fueled lamps, lubricated machinery, comprised candles. Sperm whales were a favorite quarry, and the center of the whaling world was New Bedford and Nantucket in Massachusetts. In 1711, before commercial whaling, scientists' best estimate is that 2 million sperm whales swam the ocean. A whaling blitz in the 1840s took ten thousand whales a year, cutting their numbers to around a million and a half. A lull ended with the invention of mechanized harpoons, and from the 1940s through 1970s, until the International Whaling Commission moratorium, whaling killed 35,000 sperm whales each *year*. Approximately 800,000 survive today, about less than half of pre-whaling numbers.

Sperm whales are strikingly odd. A third of their body is their distinctive megalithic head, behind which their bodies are striated with wrinkles. Sperm whales live around the world wherever ice-free water reaches a kilometer or more in depth. They plunge deep to hunt creatures in those dark layers, particularly squid.

Herman Melville's 1851 classic *Moby-Dick* describes whale physiology with the familiarity of a naturalist and butcher combined. In one memorable scene a crewman falls into the sperm-filled head of the whale. Melville doesn't believe the whale can hear. "The ear has no external leaf whatever; and into the hole itself you can hardly insert a quill, so wondrously minute is it." And though he describes the whale's clicking apparatus with astounding detail, Melville claims "the whale has no voice, unless you insult him by saying, that when he so strangely rumbles, he talks through his nose. But then again, what has the whale to say? Seldom have I known any profound being that had anything to say to this world, unless forced to stammer out something by way of getting a living. Oh! Happy that the world is such an excellent listener." Shipboard surgeon Thomas Beale's 1839 *Natural History of the Sperm Whale* also describes the sperm whale as a quiet creature.

Frederick Bennett's *Narrative of a Whaling Voyage Round the Globe* describes, "When the whale descends to any considerable depth, a sound, which may be compared to the creaking of new leather, is conducted from its body along the line . . ." One of the last whalers in New Bedford, Henry Mandley Jr. had related to William Schevill that when the sea was calm, he'd sometimes heard "impulsive noises from below" near sperm whales. He'd called it, the whale "snapping his spouters."

In March 1957, Schevill and L.V. (Val) Worthington were in the vessel *Atlantis,* the Woods Hole ship that had helped the USS *Semmes* untangle the underwater paths of sound with "sound school," when they crept within 15 meters of a sperm whale pod. They started recording and captured *three* different sounds: a percussive sequence of thuds; a "grating sort of groan" that sounded like a rusty hinge creaking; and numerous clicks. The last was the most common, and "loud enough to blacken the sounding recorder paper." Schevill speculated that Mandley's "snapping" was the loud clicks he heard on *Atlantis.* And "Bennett's creaking reminds one of Worthington's groan."

It would be almost ten years, in 1966, until Schevill and Richard Backus published details of sperm-whale sounds they'd recorded in autumn 1958. Sailing between Savannah, Georgia, and Cape Cod in more than 2 km of water off Cape Hatteras, they heard sperm whales clicking and recorded the rhythms, speculated the whale was echolocating, and estimated the range of its sonar.

At Tavolga's 1963 Marine Bio-Acoustics conference this data helped identify the sperm whale as the source of these percussive sounds. Ridgway writes of Schevill and Backus's work: "Their astute observations revealed many of the workings of an animal of which I was completely unaware."

To read the pitch-black depths in which they hunt, the sperm whale must actively use sound—echolocate with clicks. Just listening won't work because its squid prey is silent. Yet clicks don't travel very far underwater. A click at the surface has no hope of reaching prey. The sperm whale must make a click that is more powerful and closer to its prey.

To get closer, the whale dives into pressures that would crush a lesser creature. A kilometer down the pressure is 100 times that of the atmosphere, collapsing human rib cages and compressing lungs to less than 1 percent of their surface volume. The whales' heartbeat slows to use less oxygen.

And then, the sperm whale must make a truly epic click. Two organs nestle inside its huge nose, one above the other, like two massive pieces of candy corn. The upper is the spermaceti organ, its narrow tip at the top of the whale's nose, blunt end at the rear against the skull. Beneath the spermaceti organ lies an oily material called "junk," homologous to melons.

Other whales' blowholes are comprised of both nostrils. The sperm whale's is a single nostril, skewed left, at the tip of its nose. Inside are the monkey lips, with an air sac in front of them, reflecting sound backward. When the lips snap, most of the click reflects back through the spermaceti organ to the skull and its air sacs, which then reflect the pulse forward and out through the lower junk organ. This double reflection focuses the click to a 10-degree arc in front of the whale, at up to a stunning 235 decibels. Ranging from 100 Hz to 30 kHz, this click can reflect off a squid at 500 meters.

Satellite tags on sperm whales as they went about their day showed them foraging about 75 percent of the time. They dove between 600 to 1,000 meters down, for an average of forty-five minutes, resting briefly between dives. They clicked their loud, slow clicks throughout the dive but didn't buzz until they were quite deep, and had zeroed in on their prey.

Sperm whales may be the loudest clickers, but they aren't the deepest divers. The beaked whales can dive up to 3,000 meters—3 kilometers—so deep that if they surface too quickly, they get the bends. They don't start their click trains until they reach great depth. They slurp up their prey, and where they hunt the ocean floor is gouged with marks left by their jaws.

ON THE INLET, smooth gray backs rise and fall as the afternoon wears on and we wait for the tide to turn. Running beneath the thrill of hanging out with wild whales is a thread of danger, and I think of the mud so close beneath us, like shifting dunes.

Small boats are known to founder here. If you happen to strand in the muck, like an unlucky or disoriented beluga, you may not just be stuck. As the water retreats and pulls the boat along the runnels, the boat may end up on very-not-flat mud. If this happens, the boat tips, spilling its inhabitants into freezing quicksand.

Yet beneath and around us, the belugas click and buzz and read their surroundings with apparent ease.

Suddenly, the pool's mood shifts. The aimless playing becomes a slow migration toward the big white adults, whose backs rise and fall away from us. "The tide is all," Paul says, squinting out at the surface. New bubbles rise around us. The boat rocks with waves, rolling on beluga backs as they scatter outward. To the most careful eye, nothing has changed, but I don't really need to glance at the clock on my phone to confirm that the tide has turned.

CHAPTER 6

This is Me: How Sounds Define Identity

The lights begin to twinkle from the rocks:
The long day wanes: the slow moon climbs: the deep
Moans round with many voices. Come, my friends,
'Tis not too late to seek a newer world.

—*Ulysses,* as translated by Alfred, Lord Tennyson

BACK ON COOK INLET, A few hours before the tide turns, we drift in the zodiac. Anchorage is an hour behind us, and the inlet with its mantle of mountains encircles us. We are, by human standards, isolated. But I don't feel alone. These young belugas romping around us, even beneath the water, are arousing what social sense I have. I feel like I am at a party. These animals are *socializing.* Whales and dolphins are mammals, and they brought their complex social bonds into the sea.

Belugas' social structure is still mysterious, but they seem to form so-called fission-fusion groups, wherein smaller knots of friends or family often come together in groups up to a thousand strong, or split off into their constituent smaller units, such as groups of young males or mothers with their calves.

The sun lowers. The calm water is liquid gold. We are killing time during low tide, hanging out at this party, waiting for the inlet's huge tides to turn so we can safely pull our small boat out of the water without miring the truck in mud at the lower reaches of the launch ramp. Paul Wade flicks the zodiac's depth finder off and on. Like Libby's, or that of any ocean-going small vessel, the little electronic device sends high-frequency sound pings and reads the distance and shape of the bottom from the echoes.

On the little screen mounted beside the zodiac's helm scrolls a grainy picture of what it "sees"—another instance of how we visual humans like to translate sounds into pictures. There's a blue expanse (the water beneath us) and a wavy line of fluorescent pink (the bottom), 2 meters below us. The water is empty, empty, empty—and then it isn't. A shape scrolls across the screen as we drift over it, or it drifts beneath us.

It's a pixelated beluga, like a character from an old-school videogame. Its body is in a relaxed slump, head up, back humped, its lower body and tail trailing behind. I peer over the gunwale and see only brown water, yet apparently a beluga whale is *right there*.

And there's something else. A squiggle nestled beneath the whale's chest and belly. Wade says it's likely a calf; Mom has brought her infant to the pool.

They are swimming in echelon formation, the calf midway down its mom's body, tucked against her flank. Whales and dolphins do this because it's hydrodynamically easier for the calf to swim, as though it's riding her bow wave. This formation also keeps the calf close and touching. Here, amid a riot of social interaction, a quieter bond is scribbled in sonar. I recognize with a mild shock a fundamental mammalian relationship.

Wade turns off the depth finder until they've drifted off, so its sound doesn't bother them.

How different a whale mother's experience is from a human's. Whales cannot hold their babies, only swim while touching them. Yet the mother needs to hunt to eat. At some point, as she turns her body this way or that, they will lose physical contact, if only for a second. And what then? A newborn beluga is completely dependent on its mom. It nurses for up to two years. If it gets separated from her, and can't find her again, it could starve, maybe strand, or be gobbled up by sharks or larger whales.

Mammals, as a group, have these social bonds between mother and offspring, and often within larger family groups, for safety and survival, for nurturing. Underwater, mammals lean heavily on sound to navigate or hunt, so of course sound also mediates their interactions including shifting friendships, dietary choices, and parental embraces.

Whale sounds are still very mysterious to science, and they often seem unfathomably complex. And perhaps they are, in ways we don't yet appreciate. But where research has started to unravel a few of these sounds' purposes, many seem to relate to the fundamental concerns of any social animal: *Who are you, and where are you?*

VALERIA VERGARA WAS born in Buenos Aires, Argentina, a city of millions of people, but as a girl she was obsessed with the natural world, particularly with animals' social behavior, communication, and cognition. She came to Canada in 1989 for school at Trent University (in none other than my own hometown, Peterborough). For her undergraduate thesis she studied coyotes, and for her master's she studied red foxes, putting an ad in the newspaper asking farmers to get in touch if they had foxes and then studying their social behavior at their den sites in the deciduous Carolingian forest—during the same years where, a short drive away, my brother and I were discovering sound didn't work so well for underwater trucks. Between degrees she studied mammals all over the world including baboons in Africa and even humpback whales in Newfoundland.

She had always been fascinated by marine mammals and had wanted to study cetaceans since taking an undergraduate course on their sophisticated social lives and complex communication. In 2002 she came to the Vancouver Aquarium, which had a Marine Mammal Research Lab and several belugas. One female beluga, fifteen-year-old Aurora, gave birth to a calf at the aquarium that summer, and Vergara decided she would study how a young beluga learned to make its astounding array of sounds, from birth through the first few years of life. They start learning sounds just like human toddlers, slowly developing their calls. Vergara didn't study echolocation clicks, but focused on how the belugas developed their communication repertoire, with calls in three broad categories: whistles; pulsed calls and pulse trains; and mixed calls.

Small, gray, and still bearing creases, or fetal folds, in his skin from being curled up in his mother's womb, Tuvaq made his first sounds before he was an hour old: quiet, broadband trains of pulses. He made his first rudimentary whistle at thirteen days. As he grew, his pulses quickened, up to 300 per second, at higher frequencies. His whistles rose in frequency too.

"They just gargle," Vergara says fondly of the young beluga.

Babbling isn't just a human-baby thing—young monkeys and other mammals babble too. It helps them practice vocalizing and increases the chances that someone will pay attention to them, giving them social interaction. The underdeveloped, unpolished, unstereotyped sounds of the baby beluga may be just this: babbling.

Vergara has elegant features, curly brown hair, and smiles easily. Her enthusiastic voice is frequently softened by her amazement at the animals she studies, amazement that seems just as strong now as it was in her student days. She describes how at the aquarium she watched and listened to the belugas constantly. It was tricky, as with fish, to tell which animal actually made which sound, but working with a young calf offers a delightful if temporary, solution to this problem. "Calves don't learn until they're older to close their blowholes when they speak," she says. "So, they leave a trail of bubbles." That, and they could usually pick out Tuvaq's sounds depending on where he was in the pool enclosure, and the volume of his calls on the hydrophones.

Tuvaq's sounds started off quietly. "After a week, they start becoming louder," Vergara says. "But that initial week is key; [mom and calf] need to stay together. And the way they have to stay together is auditory." Though they likely watch each other, too, sound is especially important, and so-called contact calls between mom and calf help keep them together. Underwater, where sight, touch, and smell are so diminished, the calls can be a lifeline.

When Tuvaq was born, Aurora began to make a call that she hadn't been recorded making before. In the two hours after his birth, swimming with him in echelon formation, she made this new call 588 times. The next day, she began making two variants of the same call, a broadband pulsed call now known to be a contact call. From her son's birth until the time that the other belugas rejoined the pool three months later, 97 percent of all her sounds were contact calls. She made them whenever the two were separated, whenever there were divers in the tank, and whenever Tuvaq made any sound to her. Sometimes, for medical checkups or other reasons, aquarium staff had to separate the calf in a small side pool, prompting a flurry of calls.

At four months, Tuvaq started combining pulses and whistles, and began to imitate this call, which sounded a bit like a creaky door.

Sometimes he would make sounds to Aurora, and she would respond with her contact call. Other times, she would make the call to him, and would respond, slowly learning the call. By twenty months old, Tuvaq could produce a fully developed, or crystallized, contact call.

Contact calls are not unique to whales, and most social animals from birds to monkeys to rodents make them. Their function is to keep in touch or keep the group together, whether that's mates, families, or larger groups. The exact information sent in a contact call varies with the social structure of the animals using it. Such calls can be individually distinct or can ID groups and families. Some species of monogamous birds use contact calls to find each other in crowded rookeries. Some monkey species use contact calls to stay in touch through the forest canopy. And cetaceans, including belugas, use contact calls to stay together in the water where vision is unreliable and the threat of separation looms—especially for mothers and calves.

When Tuvaq was eighteen months old, he met his father, Imaq. Imaq had a distinctive trill in his pulsed calls, and a few months afterward, Tuvaq began to trill too. He may have had a genetic predisposition to his father's call; or it developed with maturity—but maybe he learned it, further suggesting that belugas' calls are learned, not hardwired in.

Tuvaq became vigorous, vocal, and, as described affectionately by the staff, a "brat." Then, one day in 2005, after a routine medical checkup, he ate some herring, and went out to play in the pool. He suddenly stopped breathing and died.

An autopsy gave his cause of death as a congenital heart defect.

By this time, Vergara had begun to shift her work to belugas in the wild. She was in a tent at the Nelson River Estuary when she heard the news via satellite phone.

Vergara was devastated. She continued to study the contact calls in recordings of the aquarium's small group of belugas, including two more calves. The data grew, demonstrating that these contact calls functioned as a *where are you, here I am*. Belugas made them in response to separation, apparent threats, births, or deaths.

Contact calls can be easily picked out of the vast array of other sounds that make up belugas' astounding vocal repertoire. While some species make very distinct and stereotyped clicks, whistles, and pulses, and beluga calls can be distinct types, too, their calls can also shade into each other, in graded call types that combine these sound types. It's hard to group calls without knowing what the belugas use each sound for. Defining one of the calls, a contact call, and outlining when the whales used it, meant the call could be studied. But as early echolocation researchers knew, wild whales might make different calls than captive. So Vergara wanted to find out if belugas used this contact call . . . in the wild.

VERGARA "PESTERED" CANADIAN government scientist Pierre Richard, with the Department of Fisheries and Oceans, and technician Jack Orr, who were running a beluga-tagging program in the Nelson River Estuary on the shore of Hudson Bay. She asked to come up to the research camp on an island in the estuary, where the team was tagging wild belugas, to study their movements. To tag the aninmals, they would temporarily separate a beluga from the group using a net. If the wild belugas behaved like the Aquarium belugas, they would likely make a flurry of contact calls during this separation, and Vergara thought she might be able to record them.

Sure enough, Vergara observed and recorded with a hydrophone, a juvenile female beluga, briefly separated from her group for tagging. When Vergara reviewed the recordings, she was excited to hear the beluga make a series of simple contact calls. But she still needed more data.

Then whale researcher Robert Michaud visited the aquarium from Quebec. Michaud was president of the Group for Education and Research on Marine Mammals, or GREMM, which studied whales of the St. Lawrence Estuary, especially belugas: the same population Schevill and Lawrence had recorded. Michaud had studied them, and recorded their sounds, for thirty years. He had met Vergara several years before and they had discussed collaboration. Now, Vergara wondered if she could pick out contact calls from his recordings.

Michaud sent Vergara some sound files, and she listened through. One day, she heard a clear series of contact calls, and excitedly told Micahud they sounded just like the calls Aurora had made at the Vancouver Aqaurium.

Michaud then told Vergara these particular calls were recorded during an unfortunate event in 1999, in which a calf in the St. Lawrence had died and a female beluga, likely its mother, had been seen swimming about the calf's body and pushing it. This behavior may seem odd, but it's well documented in whales, and probably happens because when calves are in distress, whale parents instinctively try to help them to the surface to make it easier for them to breathe. Perhaps the most famous, if heartbreaking, example was when a mother killer whale near Vancouver pushed her dead calf about the waters off Vancouver, British Columbia, for seventeen days in 2019.

The bereaved beluga mother in the St. Lawrence appeared to have made the contact calls as she pushed the calf. If the function of the contact call is to stay in contact—*I am me, and I am here*—this was more evidence wild belugas used it to maintain a primary social bond.

Belugas are not the only whales to share contact calls. Many other cetaceans do so, too. And one species, the bottlenose dolphin, is known to have truly remarkable contact calls; they are unique to each animal.

LAELA SAYIGH PULLS up a grid of 269 purple squares. Each square contains a unique squiggling fluorescent-teal line. Some are softly humped, like a kid's drawing of hills. Some squiggles are jagged, some simple and straightish, others with periodic breaks and cuts. Some are upswept, some downswept. These contours are spectrograms—time on the x-axis, frequency on the y—of the signature whistles of 269 wild dolphins that Sayigh has studied off the coast of Sarasota, Florida, since 1986. "It is gorgeous, I know." Sayigh points out one whistle. "This animal, here, seems to introduce burst pulses in parts of his whistle." In another, where the line seems to break and then jump to a different frequency; "She makes these steps in her whistle, they're really cool-sounding." To Sayigh these are not just whistles. "They're like faces to me," she says in her clear, gentle voice. "And each one is a name."

Sayigh is a research specialist with the Woods Hole Oceanographic Institute and an Associate Professor at Hampshire College. She's worked with the dolphins in Sarasota for decades. The study has been going on since 1970 and is the world's longest-running study on a wild cetacean population. Sayigh has up to six living generations of dolphins being recorded in this natural laboratory, and more generations dating back to the project's beginning, each by name.

A signature whistle is a distinctive sound a dolphin makes that is unique to that animal. These whistles can rise and fall, can be a single sound or a short repetitive sequence, can be smooth or more jagged. A call can be jazzed up with burst pulses, changes in amplitude, or even two sound types at the same time—all the features that Sayigh's spectrograms show. And they are unique. "They're the only signal that's, at least in some ways, like human names among animals," Sayigh says.

IN 1953, FRANK ESSAPIAN (who worked at Marine Studios, later renamed Marineland of Florida, with Margaret Tavolga) suggested the resident dolphins had unique whistles. In the 1960s, biologists Melba and David Caldwell picked up that hypothesis, isolating dozens of dolphins one by one, recording their whistles, and comparing them. Like Rodney Rountree's fish auditions, this overcame the challenge of telling which animal actually made a call. Dolphins don't open their mouths when they whistle, and because sound travels so efficiently underwater, and dolphins may whistle loudly or quietly, humans can't use cues like moving mouths, sound direction, or volume to pinpoint the whistler, the cues we'd use to figure out who is talking in a crowded room.

By doing this, the Caldwells showed that each dolphin had a signature whistle, which they also used as a contact call. In the wild Sarasota dolphins, about 50 percent of the whistles a dolphin makes are its signature, but if you isolate a captive dolphin so that it can't hear or see other dolphins, that proportion skyrockets to about 90 percent, like a lost hiker yelling, "Here I am!" Mom and baby dolphins frequently exchange their signature whistles back and forth. A whistle is a higher frequency, about 5–25 kHz, which is in and beyond the higher end of human hearing range.

Each dolphin develops their whistle in the first few months of life. The Caldwells describe that at first the whistle garbles and wavers, but it finally crystallizes into something unique to that dolphin. Sayigh says that it's not clear what influences the whistle's form, but after decades studying the Sarasota dolphins as they grow and have calves of their own, she's seen some patterns. Some 30 percent of dolphins' whistles resemble their mother's. This might have to do with mom's sociability: If she is more of an introvert and often alone, then baby hears her whistle the most, and copies. A handful of dolphins whistle similar to their siblings. But most dolphins usually form a whistle that's distinct in some way from other dolphins. The whistle doesn't usually change much as the dolphin ages, though Sayigh's data set shows some exceptions.

She pulls up a spectrogram sequence of one dolphin's whistle over the years. Even my untrained eye can see the rounded peaks of her tones obviously sharpen over time. We don't know if these changes are typical, but it's like her "name" stopped rising and falling softly and became sharper. Sayigh also says when young male dolphins form packs, their whistles can start to resemble each other's.

It may seem like signature whistles are just fancy contact calls. But the real difference is that rather than being hardwired in, as many terrestrial mammals' calls seem to be, dolphins *learn* these whistles. And that is really rare. "Most animals don't learn their sounds," Sayigh says. This is a feat called *vocal production learning,* and it's unique to a variety of marine mammals such as dolphins, belugas and humpback whales; a few terrestrial animals such as songbirds, possibly bats and elephants, and that's about it, as far as we know. It's incredibly rare, and studying this ability may help us study how vocal learning and perhaps even human language may evolve.

Vocal learning happens when an animal changes what sound it makes based on something new it's learned, and, Sayigh says, falls into two categories: contextual and production. The first kind, contextual, is fairly common and happens when an animal instinctively knows how to make a sound—say, a baby monkey that is born knowing a hardwired "eagle alarm" call—but not when or how to make it. It learns, from context. If it calls at a falling leaf, and no adult monkeys react, the baby learns not to call at leaves. If it calls when there's actually an eagle circling, and the adults respond, the monkey has learned from context how to use the sound.

But it hasn't learned *the sound itself.* What dolphins, and humans, do is vocal *production* learning, which is when the animal *learns* to modify its sounds, or make entirely new ones, from listening to the vocalizations of others or to other cues. A very wide definition of production learning might include an animal raising the pitch of its calls around noisy ships, but a stricter definition is when it learns a sound it otherwise wouldn't know—mimicking human voices, or each other. The difference between contextual and production learning is a bit like the difference between a dog learning to bark on command and a dog learning to say its name in English.

Because each dolphin whistle is unique, it cannot have been hardwired in. And the occasional resemblance to the whistle of their mother or those of other dolphins shows that it is learned and developed.

There's even evidence that dolphins can use each other's signature whistles. One can make *another* dolphin's signature whistle, as though calling to it, and it can also respond to the sound of its own whistle when it's coming from another dolphin. They can use the whistles as labels. This suggests they can use *referential signals*—some signal to refer to something else.

What's the point of a signature whistle? Is it to recognize each other? Most animals have some unique qualities, like scent, form or face, or voice. They don't need names—they recognize each other by this so-called *by-product distinctiveness*. Take voice tone, Sayigh says. Each human has a unique timbre of voice, regardless of the words we use. "You and I can say the same word," she says, "and someone who knows both of us is going to know which is the one I said and which is the one you said." Most animals seem to recognize each other without needing a name, and haven't evolved learned signature sounds. It's not clear why dolphins are different. But even the remarkable whistles are concerned with *who*, and so, *where*.

Other whales exhibit vocal production learning. And though dolphins are the only cetaceans we know with signature nametags in their tool kit, Valeria Vergara's work has offered tantalizing glimpses that maybe there's something more complex than we thought going on with beluga contact calls.

VERGARA WAS PERCHED in a makeshift metal tower, built several meters above a sandy delta on Cumberland Sound, on Somerset Island in the Canadian Arctic. It was the summer of 2014. She'd been dropped off nearby and had only a small yurt as a home for the next few weeks. Each day, she had to time her walks to the tower in deference to the tides; its position on low ground made it an island at high tide. (Only once did she mess up and get stranded.)

The tower at the water's edge gave her an eagle-eye view down into the water, where she could deploy a hydrophone. This particular inlet was ice-free and boat-free in the summer, there was almost always only dim ambient sound—and beluga whales. It was an ideal place to listen to and study the wild belugas' contact call. At this spot, during low tide, some belugas became naturally entrapped in pools and channels. Just as she knew belugas temporarily restrained at the aquarium made a lot of contact calls, if these wild animals were going to do the same, it would be here.

Sure enough, she heard clear contact calls almost instantly. When the low tides naturally entrapped some belugas, 61 percent of the calls the belugas made were contact calls, while just 10 percent of calls in the free-swimming belugas were contact calls. That made sense—they were trying to maintain contact when cut off. But Vergara also heard something else, acoustic features layered on top of the basic contact call. These features varied, and while she couldn't pair a whale with its vocalization in the wild, she did group all the contact calls she recorded during these tidal entrapments, according to these variations.

She found there were never more variations on the calls than there were whales in an entrapment. Indeed, the number of call variations closely tracked the number of juvenile and adult animals. She hypothesized that these subtly different calls, whether learned or not, were individually distinct, like signature whistles in dolphins. If so, these tags were a similar type of vocal production learning, and the belugas would be one of only a handful of species that had these "name tags." Humans do. Bottlenose dolphins do, and are the best studied, as Sayigh's work shows. And there is preliminary evidence that some other dolphin species do this, too.

Vergara muses about why these cetaceans might have name tags.

"They mingle, they come and go, they're incredibly sociable," Vergara says. If dolphins with their fission-fusion sociality—small units that frequently combined into larger groups or split up into smaller ones again—have signature calls, their social cousins the belugas are also good candidates for individual sonic name tags, as a kind of identifier in a busy social world.

Vergara realized the tower setup was a great way to study wild beluga vocals, and it could help her study another of the belugas' many sounds, this time in the St. Lawrence River population, where Robert Michaud worked.

In the summers of 2017 and 2018, she had built a tower at Baie Sainte Margeurite, a beluga nursery some way up the Saguenay River, and listened to the belugas' calls there. In 2021, she built a similar tower on Kamouraska Island, just opposite GREMM's center, on the St. Lawrence's south shore, for a new project with her students, under way through 2024. The St. Lawrence belugas cluster loosely around three main areas: Baie Sainte Marguerite, Cacouna, and Kamouraska. Vergara is trying to figure out if the contact calls of each of these groups are complex in another way, identifying not the individual but the group: so-called *dialects*. It's possible, because dialects have been long known in another odontocete—the killer whale.

KILLER WHALES LIVE in all the oceans of the world, except the high Arctic and the hottest tropics. They're beefy black toothed whales with a bold white patch around their eye, a fainter gray saddle patch across their back, and broad white bellies. Their iconic forms are familiar from Sea World's Shamu to *Free Willy*'s Keiko to Tilikum, the subject of the documentary *Blackfish*. Like the other toothed whales, killer whales are echolocators and meat-eaters, hunting seals, other whales and their calves, and fish.

Coastal British Columbians like me see killer whales every now and then, if not from the shore then from the ferries that are a part of life here. The first time I saw wild killer whales, I was on the deck of a car ferry weaving through the Gulf Islands, and we were coming through Active Pass, a narrow passage where the current churns up water, nutrients, and fish-y whale food. It was a typical BC ferry trip: Clusters of visiting friends and family pointed at the blue water, the green humped islands, and swirling ribbons of brown bull kelp in the currents. A jovial young man in mirrored sunglasses and a tank top was making his way through the spectators at the railing, asking anyone if they'd like a six-pack of beer that he wouldn't be able to drink before he got on the plane in Vancouver. And then we reached the mouth of the pass, and there they were in the swirl of current: gorging en masse, the slick oil-black curve of a flank or the flash of a milk-white eye patch, twisting and diving, the huge unmistakable dorsal fins. Suddenly, we ferry passengers were united in our delight, staring at each other with wordless grins. *Did you see that?*

Killer whale society, with its groups, clans, and pods, can get quite confusing, but to understand what is known about their dialects, we need to have a quick understanding, and luckily these whales around Vancouver Island are the best-studied killer whales in the world.

Of the 50,000 killer whales worldwide, scientists estimate about 5,000 live in the North Pacific. Killer whales are matrilineal, and social groups are based on kinship with mothers (note that's not quite the same as "matriarchal," which refers to women holding power), so the most basic grouping is a *matriline*—a family of a mother whale, her children, and her daughter's children. This is generally about five to eight whales, though it can have many more. Groups of related matrilines are *pods*, and they usually share a recent female ancestor. Groups of pods that share at least one call type in their repertoire are called *clans*. Though they're usually related, albeit more distantly than pods, they're defined by similar sounds. Pods and clans that hang out with each other, even if they are not all related or don't share any call types, are *communities*. And groups of communities that share similar lifestyles, diet, and so on are called *ecotypes*. Separate ecotypes sound different, hunt different prey, and don't breed with each other.

There are three "ecotypes" in the North Pacific: offshore, transient, and resident. Much of the difference in ecotypes has to do with what each group eats.

Not much is known about the offshore ecotype except that they stay in the open ocean and seem to feed at least in part on sharks.

Transient killer whales, also known as Bigg's killer whales, travel, forage, and hunt in both coastal areas and the open ocean. These are mammal-eaters, chowing down on seals, sea lions, porpoises, and (small) whales.

Resident whales, in contrast, stay near the coast and prefer fish, specifically salmon. Some resident whales live off Alaska, and two more communities live further south off British Columbia, Washington, and Oregon. The latter are the Northern Residents (of A, G, and R clans, about 300 whales) and the Southern Residents (one clan, J, just under 80 whales, and critically endangered). The letter labels for the clans originated with Michael Bigg, the first researcher to study them in detail in the wild. The names have stuck and become commonly known.

Both resident communities move up and down the coast and eat adult Chinook salmon as they return from the sea to spawn in the coast's many streams and rivers.

All of this means that in the waters around Vancouver Island and Puget Sound, three different communities—transients, Southern Residents, and Northern Residents—from two different ecotypes reside together. Live in these parts, and you start to refer to them, affectionately, by these numbers and monikers, familiar neighbors and local celebrities all in one. To the untrained eye, they look the same. Yet they do not mingle, and the reason was only unraveled with years of painstaking fieldwork—and studying their sounds.

AROUND 1970, MARINE biologist Michael Bigg noticed that he could tell individual whales apart by the shape of their gray saddle patches and their dorsal fins when they surfaced to breathe. Over decades, Bigg and his colleagues built up a photographic catalog of snatched glimpses of fins, saddle patches, and uniquely scarred backs as they broke the surface. Over the years he kept track of who lived with whom, who gave birth to whom, and generally who was who.

In the late 1970s Bigg's student John Ford began his own six-year study cataloging the killer whales, but instead of their markings, Ford listened to their calls.

The killer whales make the same three types of sounds as dolphins, belugas, and other odontocetes: clicks, whistles, and pulsed calls. These last are trains of up to five thousand pulses of sound per second, related to clicks but creating more metallic, melodic, and varied sounds. Killer whales click to navigate and hunt, and they whistle quietly among themselves for short-range social interaction, but pulsed calls can carry tens of kilometers underwater when they celebrate a successful hunt, or travel, or come near another group. The frequency of killer whale sounds reflects their world: High-frequency sounds like whistles attenuate quickly, which may be why they're used for closer communication. Low-frequency sounds carry longer distances, but if they go too low, the long wavelengths would be attenuated in their shallow-water home, squeezed between air at the surface and ocean floor beneath. Killer whale pulsed calls are in the 1–15 kHz range, a complicated mix of one or several fundamental frequencies at the pulse rate and higher harmonics.

By the time Ford began his study, Bigg had also collected some killer whale sounds, alongside his photographic database. Each group seemed to have certain pulsed calls that came up over and over again, and that were stereotyped, or the same each time an animal made it. They suspected each group made different discrete calls.

Of course, calls can vary within a species. Maybe if the species' habitat covers a huge swath of space, the northern animals make their sounds with a twang or a lilt that doesn't show up in the southern animals. Pikas do this, as do some songbirds. But when different groups *within a single area* have different call repertoires, that suggests they have different *dialects*, and that suggests they're keeping distinct from each other on purpose.

Ford focused on the Northern Residents. He cataloged all the pulsed calls among their clans and pods, and then cross-referenced them with Bigg's photo ID to confirm who was in the area when he heard such-and-such calls. When he was done, he showed that pods had a repertoire of about a dozen calls. Some were shared between pods, but some were unique to each pod. For example: The A Pods all shared most of their repertoire. The single B Pod had six calls in common with A, plus four of its own. And when the B Pod made the calls the two clans had in common, they varied from those of the A Pod.

The Southern Resident pods, meanwhile, had completely different pulsed calls from the Northerns. They even show a "striking" inflection, where their calls drop in frequency at the end, the opposite of a lilt. "This difference is readily apparent to an untrained ear," reads Ford's preliminary report. (I can confirm this: A Parks Canada biologist once played me some pulsed calls, and among the Southern Resident pods, even I could clearly tell the difference between calls. J Pod's pulsed calls sound a bit like "yee-haws." K Pod is more of a mewing, like a kitten. And L Pod's pulsed calls are more whistle-like.)

As for the transients, their calls were different again. But so was their volume. They made fewer calls than the residents, at more varied levels, from extremely hushed to quite loud, and they moved in smaller groups.

By 1991, Ford's catalog was largely complete, with several dozen pulsed calls described. These whales, all close neighbors, were making the same types of sounds, but for some pulsed calls, using different dialects. The question was, why?

Volker Deecke is now a biologist in the UK, studying killer whales in the North Atlantic, but he did his early work with Ford off British Columbia.

Deecke saw firsthand one reason why pulsed calls, instead of whistles or clicks, seem to be the vehicles of killer whale dialects one day when he was on the water off the BC coast, following a group of transients. They had been quiet all day, but then "they turn a corner, and all of a sudden they start vocalizing loudly, and tail lobbing and breaching. And then an hour later, or two hours later, another group comes in." He then discovered that another of his colleagues, Graeme Ellis, had been following and studying the *other* group. Comparing notes later, they deduced that Ellis's whales had been about 30 kilometers away from Deecke's when "they [Ellis's whales] had made a Dall's porpoise kill and started vocalizing." This was about the time Deecke's own whales began to react and breach. "And then Graeme lost his whales, I followed mine . . . they must have decided to join up." The only way the two groups could have coordinated the meet-up was acoustically, as sound is the only thing that could have spanned 30 kilometers instantaneously. And the only sound the whales make in that range is pulsed calls. So, it makes sense that the call with the energy to travel between groups is the call with the dialects.

Deecke's work also suggests a reason why transients and residents make such different pulsed calls, at different loudness, despite sharing the same waters: food. The two "ecotypes" are differentiated by their entirely separate menus.

Resident whales hunt fish—specifically, Chinook salmon, almost exclusively. Salmon can hear some sounds, but like most non-hearing-specialist fish, their hearing isn't particularly sensitive, and they can't hear high frequencies. Transient killer whales, conversely, hunt marine mammals like seals, sea lions, and porpoises. These animals can hear well underwater, and in the same frequencies as a pulsed call—or an echolocation click. (Killer whale clicks are broadband, like the clicks of dolphins, but peak at lower frequencies, between 10 and 20 kHz.) Hunting something that can hear you as you search for it means you need to be quiet. "Echolocation is a great way to find fish," Deecke says, "but it's not a very good way to find mammals. The mammals can hear those sounds and make some very informed career choices when they do." So transients, at least while hunting, don't make many pulsed calls or even many clicks—they hunt by ear.

Deecke found that harbor seals know the difference between the pulsed-call repertoires of resident salmon-eaters and transient mammal-eaters. He played for some seals recorded calls of mammal-eating transients, and the seals freaked out and fled. But when he played the pescatarian residents' pulsed sounds, seals didn't care. He also played seals the calls of Alaskan residents they would never have heard before, and the seals again reacted strongly. It wasn't any resident they objected to, but those whose calls they had learned were harmless. Deecke discovered seals are habituated to the local fish-eating whales' calls, because they've learned the animals that make them aren't a threat. "It's crazy, initially, but it makes a lot of sense," Deecke says. It takes energy to swim away from a predator. A seal wastes a lot of energy if it flees from every whale sound if only some of them want to eat it.

But was this instinctive—or learned? Deecke's findings suggest that young seals flee from any call, but as they grow, they learn to ignore the residents, perhaps by watching which sounds the adults react to. A seal would have to learn some fifty different pulsed calls in these waters to be able to identify the safe ones. And that's likely what they're doing.

"John [Ford] had done his PhD on the differences between residents and transients." Deecke smiles. "And it turned out the seals had figured it out a long time ago."

But what about the finer divisions Ford identified, between resident pods and clans? Why might they stay so separate? Because they do seem to use these pulsed calls when they're being a *group*, not a single whale.

In 2000, Deecke studied the dialects of the Northern Resident clans and groups in detail and found that some—but not all—of the pulsed calls seemed to change over time, their frequencies drifting on the spectrogram, or the structure of some calls changing subtly. After analyzing the immense database of recordings his collaborators had built, Deecke saw two matrilines, which did not interbreed, nonetheless modified their calls in very similar ways. These calls could not be changing because of genetics or instinct. This strengthens the evidence the calls are changing because of the whales' culture, that they are somehow tied to each group's social identity. And there is at least one reason why killer whale groups, of the size that seems to have dialects, might have very strong social identities.

Killer whales live for a long time—on par with humans, in ideal circumstances. Their family bonds last throughout their lives. As young, they learn from their elders how to eat, which requires clicking and listening, and how to navigate. Killer whales are matrilineal: Each family group is led by an older female. Males largely stay with their mothers all their lives. Older females guide their family groups. Some other species do this too: Elephant matriarchs remember old migration routes in lean times, and thus increase their group's survival compared to groups led by younger, and more inexperienced animals. This is the profound evolutionary value of memory and age.

Each killer whale group is bonded not only by blood but by knowledge of how to be a whale, passed down from grandmother. And since whales do pretty much everything else acoustically, they might define their groups by their sounds too.

Deecke now studies the dialects of killer whales halfway around the world, in the North Atlantic. He is working in two field sites, the Shetland Islands and Iceland. Many Shetland whales are mammal-eaters, while the Icelandic whales have a good number of fish-eaters in their ranks who love herring. Yet the diet-based division isn't as stark between the groups, and Deecke is hoping to untangle some of the relationships. He's already stumbled on some evidence of another acoustic arms race. As in the North Pacific, mammal-eaters hunt quietly. But the herring eaters also call much less often than the North Pacific residents, especially when searching for fish schools. For the fish of choice, herring, can easily detect killer whale sounds. When they do, they empty their swim bladder (in the quickest way they can—a fart) and sink. That's because the swim bladder's original function is to control buoyancy, but it also makes an excellent sound reflector, and clicks can pick it up easily. This strategic fart shifts them deeper and makes them less reflective to sound.

Even more intriguing: Icelandic whales make a kind of low-pitched sound when they herd herring, a sound that seems to make herring bunch together. It's quite low for a killer whale—about 650 Hz—but it's the same frequency as a call made by humpback whales, halfway around the world in the North Pacific, when they, too, hunt herring. The whales—both species—might do this because bunched herring are easier to catch. If Deecke can confirm this, it's yet another example of how sound can rule life and death underwater. To answer these questions, Deecke and his colleagues are building a long-term data set, like those of Ford, Bigg, and others. He hopes one day the Atlantic whales will be cataloged and understood, just like their Pacific brethren.

Contact calls, signature whistles, and dialects are what happens when complex social mammals enter a realm where sound is the primary medium of communication. They are declaring their identity, whether individual or group, loudly or softly—but only when their prey can't hear them, or at least the prey that know better. They are warning, mourning, reassuring, locating, bonding. What exactly should we call it?

THE EXACT DEFINITION of language, frustratingly, seems to depend on whom you ask. Definitions range from "a system of communication with vocabulary and grammar" to the one provided by linguist Edward Sapir, who wrote, "Language is a purely human and noninstinctive method of communicating ideas, emotions, and desires by means of a system of voluntarily produced symbols." (Phew.)

Biologist Forrest G. Wood, who worked closely with dolphins and marine mammals at Marine Studios and Point Mugu, says in *Marine Mammals and Man:* "Language is a special form of communication by which things—and classes of things—can be named, abstract ideas can be expressed, and discussion can occur. A true language is open-ended. It has the scope, range, and flexibility to handle any concept or situation the culture might present." Or, to look at it from another angle, "The difference between nonhuman vocables and human speech, it has been said, is the difference between saying 'Ouch!' and 'Fire is hot.'"

Linguistics are beyond the scope of this book. But like many people I was curious if marine mammals' intriguing vocalizations—signature whistles, contact calls, dialects—count as language. This question, though perhaps naïve, comes from simple curiosity about how marine mammals might think. After all, our own language, our words, profoundly shape our own minds. How much do the sounds of an intelligent and vocal animal like a killer whale or dolphin or beluga shape its mind, its what-it's-like, its *umwelt*? As the philosopher Ludwig Wittgenstein said, "The limits of my language are the limits of my mind. All I know is what I have words for."

The short answer seems to be that most scientists say language requires *syntax*, where the order of words (or sounds, or elements) matters. "Bring the ball to the hoop" means something different from "bring the hoop to the ball." And the majority opinion is that marine mammals' vocalizations lack this syntax.

Even those dolphin signature whistles. Sayigh says language does require referential signals, and dolphins seem to do this by using each other's "name tags." Yet she doesn't think that it passes whatever definitional bar language has set. "I won't say dolphins have language; we don't have evidence of that at all," she says.

Deecke says dialects also don't count as "language." They don't have syntax and aren't known to transmit abstract concepts. Deecke also makes a more mathematical argument.

A "bit" is a single piece of information. The bit rate of human language is pretty high. The bit rate of whale calls is much slower. "A matrilineal line has maybe fifteen call types. It's like having an alphabet with fifteen letters," he says. "When you look at typical vocal rates, it's a call per individual per minute. If I spoke a letter to you every thirty seconds, we'd be here until probably ten years from now. It's very obvious it's not a language as we know it."

But he also says that the box we label "language" is something of a restrictive human construct. "By constantly comparing to our own frames of reference," Deecke says, "we are missing the interesting questions. What are animals doing that we are rubbish at? Those are the far more interesting questions than, can animals do what we are good at?"

And some whales are very good at something that does seem to involve syntax: song.

CHAPTER 7

Tones, Groans, and Rhythm:
The Wonder of Whale Song

Traduttore, traditore.

—Italian adage meaning "Translator, traitor," suggesting that some
of the original meaning is always lost in translation

HUMPBACK WHALES TRAVEL BACK AND forth between disparate patches of ocean because no one place gives them everything. They need safe and warm waters in which to give birth and nurse their vulnerable young. The humpbacks also need lots of food, but food draws meat-eaters like killer whales and sharks to whom a baby humpback is a delectable meal. And so humpbacks swim from food to safety and back, with the turning of the seasons. Generally, they migrate from cool, food-rich, high-latitude waters in summer to warmer, food-poor calving grounds in winter.

About fourteen distinct populations of humpback whales exist throughout the world. The western North Atlantic whales migrate from the Caribbean and Venezuela, where they breed and give birth to their calves in the winter, all the way up to the waters off Canada, where they feed in the summer. On each transit, many pass the island of Bermuda, like a familiar waypoint on a commute.

Bermuda sits on the continental shelf just before the seabed drops into the Atlantic's abyss. This proximity to deep water makes it a good site for a sofar hydrophone, one of the network that was built to listen across oceans following World War II, and the Palisades station was built in St. David's. From the shore listening station, a cable snaked out and down to a hydrophone.

In the early 1950s, Frank Watlington was listening at the station when he heard whoops, moans, and other unworldly sounds. He began to record them. In 1955, as the sounds traced through the water, he sighted humpback whales swimming within several kilometers of the hydrophone.

Watlington recorded the whale sounds for years, but he didn't tell anyone. Whaling was still quite legal, and Watlington feared whalers might flock to the humpbacks' giveaway sounds. Then, in 1967, he met young whale researchers named Roger and Katy Payne, and gave them his tapes.

After Watlington shared the recordings with the Paynes, they shared them in turn with a young acoustician named Scott McVay, who had experience with underwater sounds. As a student at Princeton in 1961, McVay heard a talk by dolphin researcher John Lilly, the very same whose book *Man and Dolphin* had intrigued the navy. McVay approached Lilly after the talk with questions, and an impressed Lilly offered him a job. But Lilly's work was about to take an odd turn. He'd always used unconventional tactics; once, he had a dolphin live in a tank with one of his assistants nonstop for ten days to observe how human and cetacean reacted to each other over extended periods. Now Lilly's methods expanded to include using LSD on both humans and cetaceans in the search for a common language. After two years, McVay returned to Princeton as an assistant.

When Roger Payne showed him the tapes, McVay was intrigued. But to really analyze them, he had to put them into another form. So he printed out the songs on the sonogram machine at the university's birdsong lab.

The analog spectrograph spooled out long black-and-white print-outs, which showed frequency of sounds over time, just like Sayigh's spectrographs depict dolphin whistles. McVay and Payne saw that the squiggles and lines that represented sweeps and moans repeated every few minutes, shifted, repeated again. There was *structure* here.

In the 1930s, biologist W. B. Broughton had proposed criteria for what counted as "song" in animals, depending on how strictly one wanted to define the term. In the strictest sense, song is "a series of notes, generally of more than one type, uttered in succession and so related as to form a recognizable sequence or pattern in time." If bird-song was "song," then Payne and McVay believed the humpback sounds more than met the same criteria.

In analyzing the song, they drew on a body of work that had already been established in terrestrial environments, in a group of species known as passerines, or songbirds. Now similarities between whales and birds emerged.

If you play birdsong slowed down, it sounds very much like whale song, and vice versa—the same mix of tones and broadband, groans and whistles. Once, at a conference, I sat with acoustician Dave Mellinger of Oregon State University, long after the day's academic sessions and cocktail parties ended, even as the cleaning crew cleared the ballroom around us. I was simply riveted as he played me birdsong slowed down, and a humpback song sped up. They were, to my ear, if not exactly the same tune, very much by the same band.

Payne and McVay released an album in 1970 through *National Geographic* entitled *Songs of the Humpback Whale*. They wanted to spread the word about whale song beyond academia, to the public, and to raise awareness of the beauty beneath the waves. It wasn't the first record album of underwater animal sounds—that was the album of St. Lawrence belugas in 1950 (and Winthrop Kellogg's two-volume *Sounds of Sea Animals*, volumes 1 and 2, released in 1952 and 1955 respectively by Folkways Records, a label that worked to record dialects and languages of peoples and animals around the world). But it was certainly the most popular.

The following summer, they published their paper "Songs of Humpback Whales" in the journal *Science*, with McVay's spectrograms as the cover. McVay's painstaking printouts evoke musical notation, and indeed David Rothenberg, in his book *Thousand Mile Song*, points out they resemble the medieval musical notation of "neumes," an archaic squiggling notation used to record religious chants that preceded modern musical notation.

The haunting sounds, and the eerie beautiful squiggles, helped kickstart public efforts to save the whales, adding to a broader conservation movement that had been gathering momentum since works like Rachel Carson's *Silent Spring*. In 1982, the International Whaling Commission proposed a moratorium on commercial whaling that remains in place today.

Fishers, whalers, and anyone else who had ever sat in a boat above a breeding ground knew humpbacks made sounds. But the power of hydrophones, albums, and spectrograms showed that these sounds were *song*, by the same standards as birdcalls were song. And that word, like "language," carries implications.

IN ANIMAL BIOLOGY, "song" is a sequence of calls. As Broughton suggests, it's defined by pattern and rhythm. The *Oxford English Diction-*

ary, meanwhile, defines song as "vocal or instrumental sounds (or both) combined in such a way as to produce beauty of form, harmony, and expression of emotion."

Both definitions imply that song doesn't convey information the way a language does.

Information can be a slippery concept, but there are mathematical theories that measure how much information something contains. Volker Deecke and others have used information theory on whale song and found the bit rate is less than 1 bit per second, orders of magnitude lower than human speech, or language. The most important information in a song might simply be that someone (or somewhale) is singing. Just as those who bemoan the trite lyrics of pop songs may be missing the point of pop songs in the first place, as Rothenberg writes, "Like human music, it may express nothing but itself."

If you have ever listened to a humpback song, you may not have caught the patterns and rhythms. But there is structure there and as Payne and McVay describe, it is as follows. The shortest unit is, well, a "unit," defined as the smallest distinct sound type—a low harmonic moan, a mid-frequency "ooo" or a broadband chirp. These units are arranged in repeating groups, which Payne and McVay dubbed "phrases." Phrases in turn are organized into themes. Themes are strung into songs. A song, sung beginning to end, might last eight to twenty minutes. When a whale finishes a song, it starts it over again, like a single on repeat. A song "bout," or one whale's recital, can last more than twenty-four hours nonstop.

So, well—why? Why would a whale go to all the trouble to sing? We don't know for sure. Perhaps, like humans, they might find it intrinsically beautiful. And one charming reason why we find song beautiful may be that we are drawn, apparently instinctively, to things that have a balance of *predictability* and *novelty*, which music intrinsically does. Song is a sequence of notes or sounds that follow a pattern (predictable) and yet can vary (novelty). In his book *Why You Like It*, Nolan Gasser writes that in music, "Sequences enable a clear and identifiable narrative by delivering 2 or more (rarely more than 3) related phrases consecutively—providing both similarity and distinction." Put simply, music scratches a fundamental pattern-recognition itch in the brain, striking a sweet spot of the familiar and the surprising all at once.

Humans, birds, and whales might create song because its rhythm and repetition, its varying notes and phrases, intrinsically satisfies this balance. But then, as the adage goes, something is always lost in translation. We may never know.

After the landmark *Science* publication, the Paynes traveled the world recording humpbacks and spreading the conservation message. Katy had majored in music and biology, and over the years, she became attuned to the songs, and she saw patterns. One: whales in the same area sang the same song each year, but over time, these songs *evolved*.

Arranging the themes in a wheel, she measured which were likely to be subbed out by which, which were adopted by which group. A deletion of a theme here, an addition of a theme there, a substitution of one phrase for another.

Others began to look into the puzzle of humpback whale-song change. In Australia, Douglas Cato and assistants listened to the humpbacks' songs for years, and Michael Noad continued listening in the late 1990s. They also recorded songs in the ocean off southern Queensland, and watched something striking unfold.

Whales on the west and east coasts of Australia sing different songs. They're separate groups, divided by Australia's landmass. In 1995, all the eastern Australian singers were crooning the same song. But in 1996, two singers out of the eighty-two recorded as the whales swam northward and then southward, sang a completely different song. When the whales arrived on the breeding grounds in 1997, more than half of the whales had switched to the new song, and three whales sang a mash-up of the old and new. By the end of 1998, all the whales recorded on the Great Barrier Reef sang the new song.

The thing was, the new song hadn't come out of the blue. It was the song of the *western* Australian humpbacks. A small number of singers had likely moved from the west to the east in 1996 and brought their song with them. The eastern whales had learned the new song. This was clearly vocal production learning, and, as evidenced by the intermediate singers, remixing. Why?

Perhaps it's a strange comparison, but in 1997, above the waves, Princess Diana died in Paris and Elton John's "Candle in the Wind '97" topped the Billboard 100 for that year. John and Bernie Taupin had written the original "Candle in the Wind" for Marilyn Monroe in 1973, issued on John's 1973 blockbuster *Goodbye Yellow Brick Road*. The '97 remake had the same tune, even the same master instruments and backing vocals, but updated lyrics playing on the same themes of a young sweetheart dead before her time and mourned by an adoring public. The same, but different.

Oddly enough, another top song that year was also a remake: Puff Daddy and Faith Hill's "I'll Be Missing You," a reimagining of "Every Breath You Take," by the Police. The original song about stalking and obsessive love was now cast as a tribute to the Notorious B.I.G., who had been murdered in March of 1997 in a drive-by shooting in Los Angeles.

In *The Cultural Lives of Whales and Dolphins*, Hal Whitehead and Luke Rendall describe one possible explanation for this song shift that echoes the reasons we humans like music itself. Whales, they say, may "like the typical, but with a little novelty thrown in." The only thing better than an old classic is an old classic reimagined, the familiar and novel combined.

Ellen Garland and colleagues tracked the evolution of humpback songs near Australia over time and over great distances, assembling song data across the Pacific from Australia to Tonga. It seemed that the songs drifted eastward. Australian whales sang a given song one year, and over the subsequent years, the song would show up in progressively eastward populations, and as the original song evolved in Australia, the subsequent evolutions and remixes would drift eastward too.

This suggests, in one of the clearest demonstrations of its kind, that song is a form of humpback whale *culture*.

"Culture," like "language," is another big word with a slippery definition, especially when we apply it to animals. Some definitions restrict culture very tightly to human products like cathedrals or symphonies or recipes. Other definitions are more liberal. We humans can be precious about such things, as for every urge to identify with or learn from nature, it seems we feel an equal reflex to hold ourselves separate and unique. For everyone who finds it beautiful to think of whale culture, another finds it ridiculous. Discussions of culture in animals seems to me at least, to illuminate our already conflicted relationship with nature: How much are we a part of it, and how much are we exceptional?

One definition of culture that makes sense to me is simply "the way we do things." Whitehead and Rendell define culture as "information or behavior—shared by community—which is acquired from conspecifics through some form of social learning." Or, elsewhere, "Culture is a set of solutions to the problems of being alive." It's a word beyond the scope of this book. But most biologists I've spoken with consider that by most definitions, whales have some form of culture. And they also largely agree that it is mediated with sound and it reaches its clearest manifestation in humpback whale song. So . . . why?

Katy Payne, along with researcher Linda Guinee, showed that the structure of the themes and phrases repeated regularly throughout humpback song, to oversimplify somewhat, *rhymed*.

A rhyme can provide a catchy phrase or a clever pun in human verse but also serves as a mnemonic device that helps the singer or narrator remember long strings of information that otherwise they could forget. If whales rhyme, it might mean it's important that they remember the songs, note for note, phrase by phrase.

But why it might be important remains a mystery. We still don't know why whales sing. As far as we know, only male humpbacks sing, and while they do "warm up" their voices on the feeding grounds, or croon en route, the whales sing most exuberantly on breeding grounds. The obvious explanation, and the one most researchers think is probably true, is that song is about mating, males showing off for females. But questions remain. There's little evidence that humpback females actually prefer singers, or that singing males get more mates.

What else might it be? Humpback researcher Jim Darling proposed in 2006 that song is about male coordination or cooperation. Researcher Eduardo Mercado has advanced the theory that based in part on the signal design of the sounds, song may be a kind of sonar.

For now, the sounds themselves offer some hints. The rhymes suggest it's designed to be remembered. But it also seems to have features that will help it carry distances. The repetition of phrases and themes hints it may be redundant, so perhaps song is supposed to travel a long way. It has a lot of contrasting frequencies, from the low to the high, so it can be picked out of ambient ocean sound.

In 1984, Katy Payne spent a week at the Oregon Zoo, listening to the vocalizations of some elephants. She hadn't heard much—but she had *felt* something low, more vibration than sound. After analyzing her tapes, she found the elephants were making infrasound, below that 20 Hz human-hearing cutoff. Standing near the elephants she was close enough to feel the long, near-field wavelengths.

Captivated, Payne traveled to Namibia, where, over the next few years, she worked with researchers to show that this infrasound easily travels 2 kilometers, and probably up to 4, far enough the elephants could use it for long-distance communication, possibly because low-frequency sound attenuates slower than high-frequency sound. Payne went on to become a celebrated elephant researcher.

Her research pivot from song to long-distance infrasound is oddly apt when considering baleen whales. Humpback whales are not the only singers on the marine charts. If toothed whales are masters of name, tribe, and dialect, then baleen whales are masters of songs. And these whales, like elephants, exploit the very low and the very far.

THE BLUE WHALE, the largest animal on Earth, can grow to more than 30 meters long, which if stood on end is the height of a ten-story building. The second-largest whale, the fin (or finback) whale, can reach 24 meters. (Sei whales come in third, and humpbacks fourth.) The baleen whales also include bowheads, right whales, gray whales, Bryde's whales, and the small minke, about 10 meters long—only a modest three-story walk-up.

Baleen whales are so huge in part because they are in the water. On land, the bigger you get, the harder it is to support your own weight. (That's why insects can fall from great heights, while elephants can't jump.) It also gets harder, the bigger you get, to consume enough food to fuel yourself. It's possible the only reason whales haven't gotten even bigger is because they can't take even bigger gulps.

For baleen whales are gulpers. Their mouths are massive, and most species' throats are pleated to expand like an accordion. Gulping and filtering is one of the most efficient ways to consume a lot of calories—with the least energy. Baleen whales must find dense clumps of food—usually clouds of krill or plankton—and engulf prey and water in one huge lunge, then filter the water back out, catching the food on their baleen like the dregs in a sieve. We know comparatively more about the toothed whales, like killer whales or dolphins, because we can feed them in captivity with fish or squid. But aquariums can't host vast tracts of ocean with the plankton that a baleen whale can gulp. So we know less about these whales' social structures, bodies, and acoustic abilities. But we do know their sounds are lower than those of odontocetes. A lot lower.

The first baleen whale song to be uncovered was not in fact the humpback's, but one of the ocean's most ubiquitous whales, and was revealed with data from the SOSUS network. In 1963, William Schevill, William Watkins, and Richard Backus showed that the ocean-pervading 20-Hz "blips" they heard weren't Soviet subs but fin whales. Fins' calls are well known among ocean acousticians because they are all over the place.

Blue whales call very low, at 18–20 Hz, on the very lowest edge of the best human hearing, albeit less frequently. Sei whales, the mysterious third-largest baleen whale, are thought to be the same. These repertoires, compared to those of toothed whales, are simple. Fin whale calls are perhaps the simplest. One is the "20-Hz" call, or "A-notes," stereotyped pulses of sound centered at 20 Hz, with a high-frequency component at 85 to 140 Hz. (Like the hum of the plainfin midshipman, the call has a fundamental frequency with several harmonics.) A-notes come in simple trains, pairs (the 20-Hz blip noticed on early hydrophones), or triplets. They also make a "B call" with narrower bandwidth, also around 18 to 20 Hz. Finally, fins have a 40-Hz downswept call, likely associated with feeding. It also has a relatively higher-pitched call around 90-150 Hz, made in association with A calls.

So what is a fin whale's *song*? It's these same units, in particular sequences. Fin males sing when they make A calls with different intercall intervals.

Sei whales are less well-known than fins and blues, but their calls are also very low frequency. One is a 1.5 second down-sweep pulse, from 82 to 34 Hz, and it might be a contact call. They're also known to make pulse trains at 3 kHz, and broadband calls from 100 to 600 Hz, also with upsweeps and down-sweeps.

Blue whales make the lowest sounds, beating out fins by a couple of hertz—and they have a bigger repertoire, though not by much. All around the world, in the Pacific, Indian, and Atlantic Oceans, their calls are different, but they all have a fundamental frequency from 10 to 40 Hz. In the eastern North Pacific, they have calls dubbed A, B, C, and D. A calls are pulsed, and B calls are tonal, and each lasts about twenty seconds. The down-sweeping D call may be a feeding or a contact call. These are the blue whales' vocabulary: Their song is a sequence of A and B calls, strung together in patterns. Blue whales make about ten different songs around the world, which in contrast to the songs of humpbacks, don't seem to shift but stay static over the decades.

These songs may be near-inaudible to the unaided human ear, and profoundly simple, but they suggest remarkable things. One is that while blue-whale song structure doesn't change much, some of their calls are changing frequency. In the mid-2000s, scientists trying to use an automatic blue-whale call detector had to shift the frequency lower each year. In 2009 Mark McDonald, working with John Hildebrand at the Scripps Institution of Oceanography in San Diego, showed that components of the B calls of blue whales had shifted downward over time in the north Pacific. Decades of recordings showed the calls were 31 percent lower at that time than in the 1960s. Further, this lowering phenomenon was happening all over the world, and even in some fin and bowhead whale populations. In 2022, Hildebrand and colleagues described how the A call is shifting downward too, at about 0.32 Hz per year.

Theories abounded. One was that blue whales as a species were getting bigger, perhaps as a rebound after whaling was outlawed, as the practice had culled the biggest individuals. Or perhaps they were calling at a different depth for some reason. Depth compresses any elastic air-filled space—like a whale's lungs and throat. A deeper whale is a smaller-throated whale, is a slightly higher-pitched whale. Yet another theory was humans. Ocean ambient noise has risen sharply as shipping has grown in recent decades. Perhaps the whales were changing their tune to be heard. Or perhaps the increased breakup of ice and calving bergs in the polar regions thanks to climate change was flooding the ocean with low-frequency sound, and the whales were compensating.

But if the issue was noise, the whales should have been calling louder, not lower, to be heard. Not all populations were growing post-whaling, yet the whales shifted their pitch worldwide. And in addition to the pitch shift, the group found slight "resets" each year, where the frequency was correcting itself closer to what it had been before, then continued to decline again. Finally, the researchers also detected changes in the rate of A-call *pulses*. The cause may be some combination of all the above factors, but the simplest explanation is that there's some selective pressure for this shift among the whales themselves, possibly a cultural factor.

One of the most famous low and loud whale sounds is a deep baleen song. In 1989, Joseph George of the U.S. Navy first detected a regularly occurring whale call with a fundamental frequency at 52 Hz on the SOSUS network in Washington state. He notified his higher-ups, who reached out to William Watkins at WHOI, who in turn tracked "52" from 1992 to 2004, as it moved between Alaska and Mexico.

The "52" in 52 Hz refers to the fundamental frequency of the whale's calls. It's not quite the case that no other whale can hear it, since as we've seen, many species' calls include energy at that frequency. But no other species have calls *centered* there. The blues and fins are based at 10 or 20 Hz. Perhaps 52 was a hybrid. Perhaps it was deformed, abandoned, or a whole new species. It was likely a male, since only male baleen whales are known to sing. It vocalized faster, higher, and more often than other whales.

The whale captured public imagination after the 2004 paper. It spurred a Leonardo DiCaprio–funded documentary, *The Loneliest Whale;* an expedition to track it; an entire song by the band BTS, "Whalien 52," think pieces; essays; and an impassioned online community. (The whale recently shifted its pitch down to about 49 Hz, and recent efforts seem to have found at least one other individual at the same frequency.)

Part of the human fascination with whale song, with lonely whales or globally coordinated song change, is the suggestion that whales' minds could teach us something about our own. That they are, in some essential way, like ours, yet new. Just as learning a new language can give insight into our mother tongue, perhaps another kind of mind may shed light on what, ultimately, a mind *is*. A mind that can conquer great distances. But what would it actually be like, to reach across seas with sound? For these profoundly low and long-range calls, like all sounds, can become weird over great distances.

BALEEN WHALES TEND to make low sounds, whether call or song, for a few reasons. For one, the whales are huge, at least relative to a human, and so are the vibrating vocal folds in their throats. A bigger surface if vibrated makes a longer and lower-frequency sound wave. The pitch may be driven by female selection for big and healthy mates, as larger whales can make lower frequencies.

Then there's their natural distribution. To get food, baleen whales must both migrate and avoid congregating in too large a group, so they might need to keep in touch over tens or even hundreds of kilometers. Ergo, there is pressure for calls either loud or low, since lower sound travels farther. In general, it's hard for animals to make a sound whose wavelength is more than a couple of times its body length. An 18-to-20-meter fin whale's 20 Hz call has a wavelength of about 75 meters, so, its sound is on the low end of what it can manage, but doable. Another consequence of the low-sound-traveling-farther thing is if you want to talk long distance, it's a good idea to spend energy on low frequencies so your message will arrive in more or less the same form as it was sent, without attenuating or fading too much.

Being in deeper, open seas makes it easier to make low sounds. In shallow water, sound spreads differently than in deep water. The upshot is that long waves don't "fit" near coasts, and they begin to reflect and reverberate. This distorts low sounds in shallow water. This is added pressure to remain in open waters, also conveniently quieter than the coasts. So far, so good: an animal that needs to call over long distances finds it easy to do so.

But the open sea brings a new set of problems for communication signals.

For one, a signal degrades the farther it travels, a problem that affects baleen whales more than other species. There's also, proportionally, more low-frequency sound in the deep ocean, because these sounds travel so well. Here we find sounds whose upper frequency components attenuated a long way back.

So, a whale may want to contrast against this low-pitched sound. This might be why even simple calls sweep slowly up or down. The whale may also want to keep it simple. Distance signals, like alarm or contact calls, tend to be simple and repeatable. There's a trade-off between complexity and redundancy, and over great distance, only the simplest signals will get through.

There's another distortion unique to long-range sound channels, called multipath. A sound channel like the sofar channel continually bends waves back toward their axis, but not every sound path, or "ray," within the channel follows a straight path. It depends on the angle it enters the channel. If the sound runs straight down the channel, it travels relatively straight—but if it enters at an angle, then it will bend in and out of the channel as it travels. Because the channel is the lowest sound speed, the straighter rays travel slowest, and over long distances, this means different rays arrive at different times.

The baleen whales strike multiple balances. They call as simply as they can but vary their tones just enough to stand out against the background. They go as low as they can, given their body size. And they repeat simple calls, so redundancy compensates for distortion. That may be why the fin whale's calls have traveled so far and why scientists can track whales that appear to be coordinating over hundreds of kilometers. Inevitably, we wonder if whales could sing across oceans.

SCIENTISTS HAVE LONG tried to discover how far underwater sound really can travel. It was in the military context that an oceanographer named Walter Munk proposed in the late 1970s an idea to a secret Cold War science group called the Jasons. The scheme would send sound quite literally around the world.

Munk was an oceanographic forecaster. He had spent the bulk of his career predicting ocean conditions of internal storms and waves, discovering currents and gyres, and had even predicted the break in the surf on the beaches of Normandy between June 5 and 7, 1944, which would allow for the D-Day landing. Now Munk proposed a truly novel way to measure ocean temperature on a large scale. Because low-frequency sound traveled so far, it might be possible to send out low frequencies that would reach receivers on the other side of oceans and thus measure the ocean's temperature from the arrival times. As different rays arrived, they would have traveled through different swaths of the ocean as they swung into and out of the deep sound channel, so from their arrival times one could calculate a detailed thermal map.

Measuring ocean temperature interested an oceanographer like Munk because people had been crunching the numbers on carbon-dioxide emissions and the broader public was coming to the realization that Earth was headed for a greenhouse effect. Much of the heat would be absorbed by the ocean, and this would affect long-range sonar accuracy. His idea could show whether the ocean was warming, and if so, where, and how quickly.

He called this "ocean acoustic tomography," after CT, or computer tomography, technology that used X-rays to visualize slices through flesh. He told the naval research brass that there was a simple one-shot experiment that could test the idea. There was a place with the apt name of Heard Island in the south Indian Ocean, positioned in such a way that from there, a straight line could be drawn upward through the Indian Ocean; eastward, past Australia and through the Pacific to Oregon; and westward, beneath the Cape of Good Hope through the Atlantic to Bermuda. It also happened to be equidistant from both Oregon and Bermuda. In other words, not only could a sound from this exact point on Earth theoretically reach the U.S. East and West Coasts at about the same time, but low-frequency sound from Heard Island could flood *every* ocean of strategic interest. And it just so happened that one of the only ships capable of making such sound was testing low-frequency sonar in the South Indian Ocean.

The Heard Island Feasibility Test, in January of 1991, succeeded, with sound reaching its targets within a few hours. Munk proposed stations set up in Point Sur, and in Hawaii, under the name "acoustic thermography." But scientists and the public raised concerns about the impacts on marine mammals. (Researcher Lindy Weilgart coined the phrase "a deaf whale is a dead whale.") The projects officially stopped transmitting in 2004.

So, humans could cross the world with sound. Could whales?

In 1971, Roger Payne and Douglas Webb analyzed fin whale calls and attempted to calculate how far they might travel in the sea. They'd noticed that while the largest groups of whales were of toothed species, baleen whales seemed to be in smaller groups, or even solitary. But were they, really? Their diet made it hard to find patches of the ocean with enough plankton to feed great herds in the numbers that dolphins and belugas gathered. But maybe, they reasoned, in lieu of a physical group these quintessentially sonic animals formed a so-called "acoustic herd." "A whale in acoustic contact with another whale," they write, "is not alone."

In today's ocean, a call spreading from a fin whale in deep water could travel some 72 kilometers. In a preindustrial ocean, sans shipping or other human-made noise, the signal could go some 725 kilometers. But that was assuming water was uniform. If whales had swum in the ocean for 50 million years and leaned on sound as heavily as they seemed to, perhaps they had learned that if they swam to certain layers, sound traveled a very great distance. Payne and Webb suggested that if the whales used the sofar channel, songs could easily travel 845 kilometers. And in yesterday's ocean, in the preindustrial silence in which they evolved? The number the researchers calculated was 18,500 kilometers. In the table summarizing all this, the final cell reads, "no ocean big enough." Though the widest part of the Pacific, between Indonesia and Colombia, is about 19,800 kilometers, the sentiment remains.

Scientists have recorded whales on opposite sides of oceans that seem to be making sounds in sync with each other. Their calls can reach hundreds of kilometers over the right circumstances. They could use the channel.

Let's consider what that would look like. Even under ideal conditions, after such a distance a sound would be very distorted. As Volker Deecke describes, it would be "this washed out reverb-y kind of sound."

Laela Sayigh mused that even though the thought was impressive, and the phenomenon possible, she wasn't sure why any animal would particularly *want* to communicate with another a thousand kilometers away, in part because the distance was so great that any information, from the standard "I'm a whale, and I'm here" to any more complicated message, would be redundant by the time even a speedy whale like the fin could get there. "The idea that whales would actually communicate across oceans doesn't actually make much sense," she says.

Deecke describes distance as a double-edged sword. "You want to reach distant animals, but sometimes you don't want everyone to decode the signal," he says. Long range is not always a good thing. "We use our communication very selectively depending on what we're trying to do," Deecke says, "and there's no doubt that other animals are doing the same thing." There's also no evidence that baleen whales habitually travel to the deep sound channel to call.

Yet it doesn't seem strange to me that, if you have a culture, and bonds, you want to hear from friends and family even if you're not visiting them.

Whether or not whales call across the oceans, understanding the how and why requires us to stretch our understanding of what a signal is, both in space and in time. Sound may be faster underwater, but it's still finite. Sending a call across the Atlantic, depending on the route, could take hours. Sending a call across the Pacific, much longer. Leaving aside travel time, even if whales just want to stay in touch, they wait minutes or hours for the signals. The lives of low-voiced whales are those of long distances, and distant compatriots, and their sounds also share frequencies with those of Earth itself.

JUST AFTER 1:30 A.M. in November, fewer places on Earth are bleaker than 300 kilometers off the coast of British Columbia, and straight down into the dark, cold Pacific Ocean. But it's not empty.

The fin whale moves through the water, a sleek gray corvette of an animal. It surfaces to breathe every few minutes, then arcs back down into the dark. Each chilly inhale of late-fall north Pacific air fills its lungs and sustains its warm mammalian body for up to a quarter hour as it swims or feeds. Or sings.

Inside its body, this island of warmth in the cold sea, the whale's throat has folds and sacs as large as a rumpled bedsheet. The whale pushes some blood-warmed breath through these sacs and folds, back and forth. Whereas a human must force air out of its body to speak, the whale need only slosh its breath back and forth in its throat. The whale sloshes in pulses: *blip-BLIP. Blip-BLIP.*

Inside his flesh is warm: outside is tough skin and chilly blubber. Nonetheless, it transmits the vibration to the water, as it's at 20 Hz.

This sound is far from the only one juddering these water molecules tonight. Nearer to the coast, seals sing, fish grunt, even crabs and shrimp click or crackle. But out here, the fin whale hears only a few other whales. For long stretches, it hears the hiss of breaking waves that everywhere permeates the ocean. Each little bubble folded into the sea's surface makes a sound when it pops. Weather never rests. Neither do the world's shipping lanes. The whale swims amid the distant growl of engines that reverberate through the sea 24/7. It hears the whisper of a mud slide and the swish of a current on rock. Even in the blackest, loneliest November depths, Eliot's pair of ragged claws scuttle across the floors of never-silent seas.

And then, more than 2,000 kilometers away, far to the northwest and deep beneath the Alaskan archipelago, a chunk of the earth's crust suddenly jerks into a new configuration. Not by a lot—humans will record this as a 5.1 quake on the Richter scale, as the seabed itself pushes the water back and forth, its shudders translating into a sound wave.

Twenty-three minutes later, the wave has distorted, scattered, reflected, refracted off bottom and surface, but still reaches the fin whale. The sound vibrates through blubber and bone up into the whale's head, into the spiral cochlea of its inner ear, where it shakes the delicate membranes and transduces sound vibrations into a nerve impulse that travels into the whale's brain. The whale *hears* the earthquake.

The sound isn't loud enough to harm its ears. But the whale is affected: It shuts up. For a few minutes, it drifts, silent in the November blackness. For a few minutes, it becomes invisible. And then, perhaps deciding that it's safe again, the whale resumes its call.

Blip-BLIP. Blip-BLIP.

This acoustic scene was caught by the Cascadia Basin node of Ocean Networks Canada's NEPTUNE network, the same network with another node in Folger Passage almost smack-dab beneath Libby's path the day we motored out to the wave-scoured seal haul-out. The network's other nodes trickle down the continental slope into the darkness. Each node has hydrophones, oceanographic instruments, video cameras, and seismometers. This last is a device that measures the vibrations of Earth's crust, picking up the distant shudders of earthquakes and volcanoes.

The Cascadia Basin Node was built on the site of a borehole drilled by the Ocean Discovery Program. Originally this hole had a CORK in it, or a Circulation Obviation Retrofit Kit, sitting 107 meters down and measuring Earth movements, pressures, and earthquakes. After all, this is a notorious subduction zone, where the Pacific floor is sliding slowly beneath North America, prone to earthquakes and scheduled for the so-called Big One either yesterday or several hundred years from now.

A seismometer picks up vibrations, even very subtle ones. And it picked up the fin whale's 20-Hz call. The song is so deep, low, and powerful that some whales such as fins and blues are often caught on seismometers. Because the whales' calls are so low frequency, in shallower water the wave interacts with the seabed, creating seismic signals. Some scientists have even co-opted fin calls as ersatz seismic imagers of subsea geology—once dubbed "fin whales of opportunity."

The low voices of the baleen whales remind me that sound is not an ephemeral thing. Yes, it is call and song and dialect and name, but it is also physical movement, vibration, energy shifting the molecules it moves through. Most of the time, it is intangible. But there are other times when sound underwater can be awesomely powerful—and harmful.

CHAPTER 8

Extremely Loud and Incredibly Close: How Noise Shrinks the World

> I am just as deaf as I am blind . . . I have found deafness to be
> a much greater handicap than blindness.
>
> —Helen Keller, Letter to J. Kerr Low, March 31, 1910

AS JONI MITCHELL LAMENTED IN the 1970s, and glam-metal rockers Cinderella later belted out at many an '80s high school dance, you don't know what you got ('til it's gone). I think the same is true of senses. If you've never lost a sense, it's hard to appreciate just how much you rely upon it.

Perhaps a human's best comparison to what it's like for underwater animals to lose touch with sound is a sudden loss of sight. This happened to me during a cold-water scuba dive course several years ago.

Floating in the 8-degree November chop, a storm bearing down, cold water kissing my wrists around the seal of my gloves, I hear the instructor yell to the six students including me, "There's no emotion in the ocean!"

It's only our third dive in the ocean instead of a pool. We bob in

Saanich Inlet just north of Victoria. One student, who's had chronic trouble adjusting the weight and buoyancy of his gear, yells that he's anxious about today's dive. "We aren't doing anything that we haven't done in the pool," the instructor says. "You can do this."

I'm nervous, but ready. I'm not claustrophobic, and I'm a good swimmer. We ready our buoyancy control devices, or BCDs, those vests that scuba divers fill with weights and air to rise or fall in the water. We descend.

I realize I am blind.

At first, I think this is the ocean itself. The top few meters are as cloudy as lemonade, and we've been warned plankton makes it hazy near the surface. I look for the others but see only a blurry dark shadow, maybe a flipper. And then nothing. I am alone in murk and sinking farther from the air. I'm a bit thrown, but still game. *Just hang on till you get down to clearer water,* I tell myself.

Then the haze shifts dark blue. We must be out of the plankton, I think, like a plane descending from clouds. Except my vision is still hazy. Down. Down. My knees bump the rocky bottom, and I'm four stories from the surface. The blurry blobs of other students waver around me.

My mask should be clear by now, but it's like I am wearing the wrong prescription glasses in a foggy sauna. Yes—fog. That must be the problem. I remember what I was taught to do if my mask fogged. It still makes me nervous, but I don't have a choice. I carefully flood the mask by pulling it back from my face. The freezing water shocks my eyes and nose. Then, holding the mask's lower edge against my cheeks, I inhale deeply from the regulator and make one big nasal exhale, forcing the water out of the top of the mask. This *should* rinse the fog.

But there's no change. I still can't see. It must be the water, or some problem with the mask itself. I start to breathe faster. Am I the only one who can't see? Surely we can't be doing our scheduled lessons and tests in these conditions.

At some point, the shadows around me begin to rise. Are we ascending? I try to remember how, but my mind, and my heart, are racing. I forget if I need to add or remove air from my BCD. I know I'm supposed to keep breathing, but it's a struggle, and I have no idea how fast I'm moving up. I couldn't see the depth gauge on my BCD even if I remembered to look.

I have no idea what's wrong, or how to fix it. Every instinct roars at me to thrash, scream, run—all of which are very bad ideas. I don't remember the distress hand signals we learned. Despite my best efforts to stay calm, panic reverts me to muscle memory and I feel my breath come faster, using up precious air. I know this is bad, but the best I can do is breathe and try not to rise too quickly. I realize this is the most physically afraid I have ever felt as an adult.

Blindness is not an acute threat. I can still breathe. I am not being hurt. Yet the disorientation makes me dangerous to myself.

Finally, I surface. I rip off the mask and see again.

At the surface, my class is concerned. My instructor examines my mask but can't find anything wrong with it. A fellow diver gives me some tea, and I head home early that day. The next week I buy a custom-fitted mask. I head to the shore, wade in, and dunk my head underwater to test it. Beach pebbles waver blue-gray, in sharp focus. Remasked, I return to the class, determined to continue, and complete my course.

A few months after the dive, I read Diane Ackerman's *The Natural History of the Senses* and learn that "surd," the root of the word "absurd," translates to "silent or noiseless." Although my scuba experience was a loss of sight, not hearing, the word rings true to me. It felt absurd. Nothing made sense, and my instincts kicked in just when it would be most dangerous to follow them. If a deafened animal underwater is anything like me with my garbled mask, I can begin to sense how quickly their world could become treacherous, even deadly, without sound.

BACK ON MAY 5, 2003, the sun is glinting off the blue-gray waters of the Salish Sea near Haro Strait, a couple dozen kilometers from the site of my scuba adventure. The Strait winds along the border between the (Canadian) Gulf Islands and the (American) San Juan Islands. The gleaming white cone of Mount Baker shines against the sky as currents swirl in blue and gray water beneath the green slopes of the islands.

That afternoon, a pod of killer whales is swimming off the southwest coast of San Juan Island. Several whale-watching boats sit a few hundred meters away, packed with excited tourists. These killer whales are some of the Southern Residents' J Pod, which in 2003 numbered some twenty members.

J Pod's matriarch, which Michael Bigg dubbed J-2, is known colloquially as "Granny." Other members of her extended family include her possible son J 1 (Ruffles) and many others. Like the rest of the Southern Residents, J Pod was diminished in the 1960s and '70s when live capture for aquariums slashed their numbers, and since then, decreasing salmon prey and chemical contaminants in their food web have kept their numbers low. Still, on this May afternoon, they're getting along fine.

But suddenly, depending on whom you ask, the whales either start to act a little more surface-active—or completely freak out.

There are a lot of witnesses. There are the whale-watchers, but also whale researcher Ken Balcomb, watching from his deck on the shore of San Juan Island, and David Bain, a whale researcher and professor at the University of Washington, observing the whales from a boat just offshore. Balcomb started videotaping, and soon a Seattle news crew arrived. The videos show J Pod becoming visibly restive.

Bain also describes, earlier in the day, hearing loud sounds through the hull of his boat. The whale-watchers, too, will later mention an annoying sound audible above the water.

The sound is eventually traced to the USS *Shoup*, a gray-painted guided-missile destroyer of the U.S. Navy. The *Shoup* was heading from its base in Washington, conducting readiness exercises as it meandered through Admiralty Inlet, then west past Victoria, BC, then turning back toward Haro Strait en route to a military testing range off Nanoose Bay near Nanaimo. U.S. Navy ships do readiness drills every few months, using sonar to detect mines or enemy submarines in narrow, restricted waterways. The *Shoup* was using a type of sonar called mid-frequency active sonar, or MFAS. MFAS was pinging that day at a fundamental frequency of 2.6 and 3.3 kHz, with strong harmonics at higher frequencies. At the source, the sonar was 235 decibels. Sonar sound, to a human ear, is a mash-up of nails on a chalkboard and old-school dial-up Internet.

Bain, now with Orca Conservancy in Seattle, later contributed to reports about the incident, and describes J Pod trying to head closer to the shore of San Juan Island, relaxing when the Shoup briefly headed west around the southern tip of Vancouver Island and away from the whales, and then reacting again when the ship came near once again to transit Haro Strait. The pod then seemed to hide behind a reef and split in half, with some fleeing north and others heading south.

In the Salish Sea, naval military sonar regularly sounds. Canada's Pacific Naval Fleet, and the ships and subs that comprise it, are based in and around Victoria. There's a U.S. naval base at Everett, Washington, and sections of the Juan de Fuca Strait and the adjacent Pacific are a designated military testing and training ground.

Yet secrecy about when and where sonar is used abounds. Militaries don't want to give away their capabilities. I once spoke with a local scientist who had recorded sonar from his organization's undersea hydrophones but said that occasionally the government will tell them to turn their hydrophones off while tests are done, or vet any underwater recordings before they're streamed online or publicized. That afternoon even before the *Shoup* entered Haro Strait, according to the naval report on the incident, the ship had been contacted by Victoria Traffic who requested they contact the Canadian Coast Guard and asked if they were using sonar, as their sounds were audible underwater.

Sonar, by nature, is usually focused in particular areas of the ocean or restricted to contained exercises. But when it does come into conflict with animals, it can be gruesome, causing everything from harassment and annoyance to trauma and even death.

Active military sonar sends out sounds, then listens for echoes. This sonar comes in many frequencies, but for simplicity's sake, there are two broad frequency ranges that come up in discussions of impacts on animals: mid-frequency and low-frequency. The tech developed during and after the World Wars was "mid-frequency" active sonar, centered at several kilohertz, and is still commonly used. In the 1970s and '80s, Munk's and others' low-frequency sound experiments, between 100 and 500 Hz, interested the navy because low sounds traveled farther. By around 2000, some ships sported up to 18 low frequency active sonar, or LFAS projectors. The animals most studied in terms of both these types of sonar's impacts are marine mammals.

ALMOST AS LONG as active sonar's been around, it's been accompanied by reports of marine mammals that have washed up dead or stranded on shore nearby, either singly or en masse. Disturbing as it can be for humans, whales and other marine mammals stranding is in fact a natural occurrence, and many strandings almost certainly have nothing to do with humans' noise. But some stranding events show patterns with respect to sonar that it's hard to imagine are coincidental. In the 1980s, the Canary Islands, 100 kilometers off northern Africa, saw a spate of strandings where one or more whales washed up on the shore above the waves and died. Twelve goose-beaked whales and one Gervais' beaked whale washed up on the island of Fuerteventura in February of 1985. The next summer, five beaked whales stranded. In November 1988, three goose-beaked and one Gervais' stranded, while pygmy sperm whales washed up on yet another island. Locals reported seeing military maneuvers offshore before and during two of these strandings.

These Canary strandings were overwhelmingly very deep-diving beaked species. And they showed trauma to their ears consistent with extreme pressure change.

For reasons we don't fully understand, beaked whales appear to panic more than other species in reaction to strange or loud sounds. Mid-frequency sonar, close by, is such a sound. A whale that panics 3 kilometers down may race for air and safety at the surface, and despite all its deep-water adaptations, ascending too fast gives them decompression sickness, or the "bends," where compressed gas bubbles in the joints and other parts of the body expand too quickly and burst tissues.

Startled by sonar, many whales try to flee. Killer whales such as J Pod, who live and hunt in the upper waters, may not risk the bends, but their immediate, and prolonged, reaction looks much as Balcomb, Bain, and the whale-watchers saw: The whales are upset, their feeding is interrupted, they're likely stressed, and the family can get separated. Such noise, and their efforts to avoid it, makes it harder for the killer whales to find and hunt already-diminished food in these waters, and if it happens a lot, the resulting calorie deficit hampers their growth, reproduction, and immunity to illness.

Opinions and findings on the 2003 Haro Strait incident were divided. Naval reports said the sonar would have been no louder than about 170 decibels for the whales, and their behavior was well within normal bounds, certainly not enough to cause injury. As for my very subjective opinion, I'm not sure it matters. I understand that sound may not physically injure the whale. But I viscerally understand that panic can. And Balcomb, Bain, and others with long field experience say that the whales were clearly distressed.

"NOISE" IS A tricky word. It's a colloquial term, but also, for acousticians, a technical one: It's unwanted sound that interferes with an acoustic signal being perceived. Noise isn't one amplitude or frequency. It depends on a situation, a species, an individual, though a sound is more likely to be a *noise* if it's very close, high in amplitude, or falls in the animal's hearing range. The mid-frequency whine of a motored pleasure boat may be "noise" to a beluga yet go unnoticed by a scallop: Its ultrasonic depth-finder may be inaudible to fish but annoy an echolocating dolphin. Some animals are even attracted to sounds that others avoid.

But the most troublesome noises often are powerful, close, and inescapable, like sonar. Unlike sonar, many are also *impulsive*.

An impulsive noise is one in which the particles go from relative stillness to their maximum amplitude very suddenly.

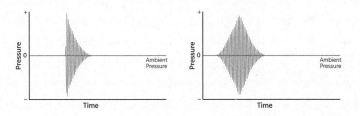

This "rise time" can make the difference as to whether a sound physically damages an animal or not. Consider you're in a powerful sports car accelerating from 0 to 60. If you accelerate slowly, you barely feel it. If you accelerate in three seconds or so, you feel the lurching thrill that is why people buy sports cars in the first place. Accelerate too fast like a fighter jet pulling too many g's (or a sports car decelerating in a crash), you knock yourself out. With impulsive sounds, particles suddenly swing to their maximum amplitude against the sensitive cilia of hair cells.

Marine mammals have a defense against this. The mammalian ear has a reflex which kicks in if a sound is very loud or sudden. A tiny muscle pulls the bones of the middle ear back from the oval window, so their energetic motion won't damage the cochlea. Most natural sounds ramp up slow enough that the reflex has time to kick in if it gets too loud. But many industrial noises, such as explosions or hammers, are intense and impulsive, rising quickly, and so create more hearing problems for animals than non-impulsive sounds of the same level.

For this reason, very loud or impulsive sounds created some of the most dramatic and best studied effects on marine life: seismic prospecting and pile driving.

Researchers modeling how underwater noise affects animals think of a source creating "zones of influence" spreading outward, like concentric spheres. The framework originated with marine mammals. For an animal in the outermost zone, where the sound is just audible, its listening space is reduced, and very faint sounds that it might otherwise hear are drowned out. A creature one zone closer might be bothered enough to change its behavior in some way, as the killer whales did that May day. Important signals are probably masked, and these behavior changes in turn affect predators or prey. Closer still, noise can cause hearing loss and in extreme cases, injury. (Sonar is one human sound that can be loud enough underwater to create all these zones.) The swinging particles of a very powerful sound are more than capable of damage to delicate hair cells.

Consider that little spiral, the cochlea. Remember a cochlea holds a coiled membrane. To simplify somewhat, the membrane, close-up, is in fact two parallel sheets. Hair cells stud the lower sheet, which undulate as the sound wave travels along its length. (See Chapter 3 for a recap on this.) The cilia, the hairs that protrude up from the hair cells, brush against the upper membrane.

However, if a sound shakes the membrane too hard, it may damage these cells. There are different types of hair cell in this complex structure, and depending on which cell type, and how severe the damage, the injury can be temporary or permanent. Temporary damage happens when the cells just get exhausted and stop firing. You've experienced this if your ears have rung after a concert. It's reversible once the cells recover. But with permanent damage the cells are hurt so badly, they can't recover. Mammal ears' hair cells do not regenerate, and the nerve to that cell atrophies.

So, the zones-of-influence model depends on proximity, ear type, and the impulsivity of the sound wave. To make things even more complex, it also depends on the individual animal. Wild creatures often have hearing issues that have nothing to do with human noise. Even the most acute-eared dolphin or humpback can grow hard of hearing. Disease, parasites, aging, or fungal infection can damage ears too.

The U.S. Marine Mammal Protection Act came into effect in 1972 and deals with, among other threats, underwater noise. This Act prohibits killing or harassing marine mammals, including with sounds. Industry or military must apply for an exemption to the Act if they're going to make noise, and since the Act uses the "zone" framework to model impacts and exemptions, the nosie maker must specify if they are going to engage in "level A harassment" or "level B harassment." Level A might happen if a sound will be very loud or close to animals, and cause death or permanent hearing loss. Level B refers to causing temporary damage, behavior changes, or reduced listening space. To get an exemption, you need to say how many animals you expect to harass, and at what level.

Predicting this is not as absurd as it may seem. Over time, scientists have built up data on where the line may be between temporary and permanent effects. Gathering this data involved many hearing tests of different species and observing how they reacted to sounds—largely focused on mammals.

The regulations have grouped animals by hearing range: low-frequency cetaceans (baleen whales, mostly affected by sound below 300 Hz); mid-frequency cetaceans (including humpbacks and killer whales), and high-frequency cetaceans like dolphins, beaked whales, or pilot whales that hear sounds above 100 kHz; and then seals and sea lions (though the precise names and definitions of these groups are shifting as more research is done). For each group, the regulations give the decibel levels that are likely to cause permanent damage, versus temporary damage or behavioral changes. The line has been about 120 decibels for continuous noise like shipping and about 160 decibels for impulsive sounds like seismic air guns, pile-driving, or intermittent sonar.

These categories are still imperfect. All baleen whales are "low-frequency cetaceans." But we haven't been able to study or test them as closely as more captive and accessible toothed whales. Instead, we've had to create models based on what we do know and can observe: ear anatomy and behavioral reactions to sounds in the wild. It is clear that not all species are the same. For instance, minkes, one of the smallest baleen whales, react to mid-frequency sonar more than their baleen brethren. Minkes also make a rather charming, almost mechanical-sounding, *boing* call that happens to be in this same mid-frequency range. Perhaps sonar resembles that call to their ears.

Baleen or toothed, a whale's reaction to a sound depends on what it's doing. If blue whales are feeding at the surface, they don't react much. If they're feeding at depth, they may dive. In a series of experiments in Norway, military sonar played to tagged sperm, pilot, and killer whales caused them to stop feeding or foraging, and in some cases to stop calling. Grouping species is better than nothing, but it almost certainly doesn't capture the nuance of their acoustic lives.

The zones are conceived for marine mammals, but we are expanding them to other animals—slowly. Invertebrates aren't yet mentioned in these regulations, though we now know many are profoundly acoustic. Then there are fish.

Mid-frequency sonar was assumed to be out of the hearing range of most fish, so marine mammals had taken precedence. But fish certainly could hear low-frequency sonar. The navy funded tests on low-frequency sonar on fish in Seneca Lake, one of the Finger Lakes in upstate New York, where a barge in about 140 meters of water sent out sonar signals.

Scientists first studied rainbow trout, a fish with no special structures to detect higher frequencies. They put the trout in a tank and watched their reaction to sonar played for five or ten minutes at 193 decibels. The fish startled and swam about, but none died. Then, they checked the fishes' ears using a high-powered microscope. A loud sound might create enough movement between otolith and cilia that they would shear off. This didn't happen, but immediately following the tests, the trout's hearing sensitivity at 400 Hz shifted upward by 20 decibels. The fish had temporary hearing loss.

The fish showed no physical harm like injury or death. They were only startled and their ears rang briefly. But their *behavior* changed. We don't know what it means to a rainbow trout's survival if it starts swimming erratically. But for example, their close relatives, the salmon, are the prey of resident killer whales. And we could imagine that a confused fish may be bad at evading a hungry whale.

But this test was just one species. Even within the cage, different fish respond differently to the sonar. Different ages, or different sexes, could react differently. Heck, as Sisneros showed, the same fish might have different hearing at different times of the year. Would a female midshipman find sonar a mere annoyance in October but an impediment to mating in May? To know what the effect truly was on a population of fish, not just a cage of individuals, meant gathering further data to untangle these questions.

Scientists subsequently tested sonar on bass, perch, and catfish (the latter of which has a swim bladder with those extensions that can boost hearing sensitivity). None showed permanent hearing changes—but they reacted to the noise. The tests were among some of the first done on creatures that were not marine mammals. But fish and especially invertebrate data are still lacking.

As sonar showed the power of sound over marine mammals, another noise showed acoustics could be downright deadly to the backboneless: underwater air guns, whose thudding, explosive sounds search the layers of rock beneath the sea for lucrative oil and gas.

IT WAS NIGHT off the northern coast of Australia in the mid-1990s, and Robert McCauley was standing on the deck of a ship, looking out over the water. The ship worked for a petroleum company, on an expedition searching for pockets of oil and gas beneath the ocean floor. To do this, it made powerful pulses of sound with seismic air guns. The intense sound penetrated the ocean floor and returned echoes from layers of rocks, sand, and oil and gas deposits. For the past two years McCauley had worked alongside such ships studying how humpback whales responded to seismic sound.

Exmouth Gulf nestles in the northwest corner of Australia. It's a well-known humpback-whale nursery, where mothers fatten their calves with gallons of yogurt-thick milk a day in preparation for migration. But the whales' path out of the gulf cuts straight across a swath of seabed that contains oil and gas deposits, and the petroleum company needed to know if the noise of their surveying air guns affected the whales. That was where McCauley came in.

"We ended up being out there on a small boat," he recalls. "The seismic vessel was doing east-west transects with track lines. And the whales were moving from northwest to southeast."

That evening, he was on board the large ship on an invitation from the crew, to take refuge from bad weather. He might not have seen what happened next had he been in the small boat, or if it hadn't been dark. As he watched the water where the seismic source was operating, he saw shimmers of light rhythmically appearing and then vanishing, spreading what seemed like hundreds of meters across the ocean's surface.

"It would send, like, a wave out from me," he says. "I couldn't see how far it went, but it went quite a distance."

Bioluminescent plankton are tiny photosynthetic algae that can produce a warning flash or glow of light when they're touched, jostled, or disturbed. They evanesce in the breaking surf, glow on the surface of "milky seas," or trail behind the agitating hand of an amazed swimmer, like the magical sparkling dust of a childhood fantasy. It's common in some parts of the sea, and McCauley thought there must be some here, and that something must be disturbing it, jostling it on a massive scale. The obvious culprit was the sound pulses of the ship's surveying equipment.

"So, I wondered what did happen to all those little animals when I did fire an air gun," he says.

This chance observation wasn't a scientific study, and McCauley wasn't allowed to publish the proprietary data from the seismic tests. But his curiosity was piqued. By the mid-1990s, he knew that seismic sounds might disturb large humpbacks, which was why he was studying them. No one had asked what these sounds did to very little creatures like plankton.

Plankton come in all shapes, body types, and body compositions.

The word "plankton" doesn't even mean one species; it just means an organism that drifts with the water, and it includes the tiniest photosynthetic organisms in the water: phytoplankton. These tiny algae fill the same role in the sea as green plants on land, in that they turn sunlight into food. Phytoplankton are then gobbled up by zooplankton like copepods, or krill, and these in turn are gobbled up by fish and marine mammals. It's become passé to refer to food "chains," which most scientists now conceive of as more like webs. Ocean food webs get their energy from phytoplankton. Algae may lack a humpback's charisma, but without them, there would be no life in the ocean, or food from the ocean for us land-dwellers.

"Plankton" also includes zooplankton. These are tiny animals like Simpson's larval fish, copepods, krill, and other animals too small to swim effectively under their own power. If the air gun could jostle these ocean denizens on such a scale, it might affect the base of the local marine food web.

McCauley says that in the grand scheme of things, mowing down a kilometer of plankton in the ocean may not be problematic, as plankton quickly regenerate. But seismic air-gun surveys often ply the same areas for months, and repeatedly scorching the base of the food web could chronically impact the ecosystem.

McCauley wanted to re-create that evening when he'd observed bioluminescence in a proper experiment, fire a seismic air gun and then measure what happened to the zooplankton and phytoplankton. But it's really hard to do research on seismic air guns in field conditions. You don't just go out and start taking measurements. The commercial ships that use them belong to a small number of industry surveying companies worldwide, and surveying is incredibly complex.

Seismic surveying for geological imaging began on land as prospectors sent vibrations into the earth and read the rocks beneath from the returning waves, revealing layers and pockets below ground through vibrations that behaved differently when traveling through rocks of various densities. When oil wells were first becoming established in the nineteenth century, seismic techniques offered a glimpse into the earth without the time and expense of drilling. But the same techniques couldn't work at sea.

Then, in the 1940s at Woods Hole, Maurice Ewing and others were testing how sound traveled. They often made these sounds in the water by undersea explosions of TNT and these sounds could travel very long distances. They noticed that as these powerful sounds ricocheted around the ocean, echoes returned from rocky layers *beneath* the seafloor. Ewing and his student J. B. Hersey mapped some structures beneath the sea-bed in Vineyard Sound using impulsive explosions, and Hersey and others continued developing technology to image these subsea layers using explosive noise during the 1950s.

Sonar used high-ish frequency sound to image subs, and animal echolocation used it to image fishy prey, but such short-wavelength, high-frequency pulses didn't have the energy to penetrate solid rock. Much as a medical ultrasound uses a longer, more powerful wavelength to go deeper inside body tissue, a longer and more powerful wavelength was needed to image rock. And because these rock formations were large, a shorter, more detail-discerning wavelength wasn't necessary. But multiple pulses were needed to build a picture.

So in lieu of powerful but hard-to-calibrate TNT, scientists developed a more controlled sound source: the underwater air gun, with a chamber of compressed air triggered to release at regular intervals. The gun released a bubble that popped loudly. Its chamber size and pressure could be calibrated for sounds of specific frequencies and amplitudes. These were combined with towed hydrophones that listened to the re-turning echoes. These echoes revealed structures that reflected sound strongly, like salt domes and hard rock caps, likely to conceal oil and gas deposits. The entire technology thus pinpointed where to drill.

Underwater seismic activity ticked up in the 1960s and '70s. By the time McCauley observed the rippling bioluminescence in the 1990s, the technology was well established, and such seismic surveying became the primary method of finding offshore reserves, which today comprise about 30 percent of global crude oil production.

Ships dragged long "streamers"—lines up to 12 kilometers long—strung with arrays of hydrophones and air guns of different calibers. They cruised at a few knots and fired the guns every few seconds. With these streamers, at slow speeds, ships typically took four hours just to turn around at the end of a pass, to avoid tangling the lines.

The average survey went something like this: First, they ran widely spaced lines, with echoes imaging up to 10 kilometers down into the seabed, zeroing in on interesting sites. Then, they would run more tightly spaced tracks at spots of interest, capturing higher-resolution images of the sub-sea geology.

But even after development of petroleum extraction infrastructure, surveys still image the same tracts of seabed with different-caliber air guns, to check the level of the deposits or find more pockets.

Trying to shoehorn a scientific study into these operations was hard. Time was money, and coordinating around a surveying ship's streamers was a nightmare. In 2007, McCauley got a last-minute chance to tag along behind an operating ship in a four-meter dinghy. He would echo-sound the plankton in the water immediately in front of the ship and then immediately after it passed, to see the difference. Like Martin Johnson and others imaging the "false bottom" in the Pacific, he'd use high-frequency sounds to reflect off the tiny creatures. But technical difficulties abounded, and while McCauley did get usable data, and it did suggest some impact, it wasn't enough to publish.

Stymied, his interest was even more piqued. In March of 2015, in collaboration with scientists at the University of Tasmania's Institute for Marine and Antarctic Sciences (IMAS), they fit a plankton study into an existing seismic project taking place in Storm Bay, off Hobart in Tasmania. This time, in addition to the sonar, McCauley and his collaborators also wanted to "bring up the bodies," as it were, of the affected plankton: here, specifically zooplankton.

They towed a pair of bongo nets, very long nets used to sample plankton. Their nets' mesh gauge varies, depending on what size of animal you want to scoop. For bigger plankton, the mesh can be fairly loose. To snatch the very smallest plankton, the mesh is so fine that the net resembles silk to the touch. Usually, scientists will tow a pair of nets of different gauges, which look like a pair of very long white bongo drums.

Storm Bay conditions included strong currents. By the time McCauley and the IMAS scientists drew up their nets behind the ship, any stunned plankton could have drifted half a kilometer, so they had to adjust for currents and boat speed. But he could effectively measure the water hundreds of meters around the boat, and he remembered how far the bioluminescent ripples had spread that night.

The researchers cruised, they sounded, and they drew up the nets from the seafloor 30 meters down, sampling before and after the ship passed. They scooped up living and dead zooplankton, which to the untrained eye look identical. So, they employed a trick to discern them. Plankton live for an hour or so after collection. On the boat, the scientists rinsed the plankton into collecting containers and immediately doused them with their favorite food, tagged with a pink stain.

"If the animal's alive, it will ingest [the food] and take it into its body," McCauley explains. This turns the animal pink. "If it's dead, it won't take it up." So, live plankton glowed pink under the microscope. Comparing living and dead plankton revealed how many had been mowed down, and how far they'd been from the ship.

The sound of the air gun knocked out plankton more than a *kilometer* around the ship. McCauley and his collaborators focused on copepods, a common zooplankton, as well as crab, barnacle, and krill larvae, collectively referred to as "nauplids." Though it was a small sample size, all the krill larvae were dead after exposure to the noise. Dead copepods doubled or even tripled after the air-gun pass, especially smaller species. The researchers' sonar imaging meanwhile showed a hole in the sea where plankton had been cleared out.

These small animals are about the same density as water. Any sound should just pass right through them: Boneless, they should move with the water as the pressure wave passes. But many plankton have oily substances on their surfaces that help them slip through the water more easily, and these substances make them respond to sound waves differently than expected. Put another way, they are less than completely acoustically transparent. Additionally, crab or other invertebrate larvae may be small, but they still have structures such as chordotonal organs, and statocysts, that let them perceive the particle motion of sound. If fish and crab larvae could feel the tickle of reef noise meters or even kilometers away, krill and copepods could certainly feel the punch of the nearby seismic guns.

Others tried to replicate McCauley's findings. In the waters off Norway, one of the most widespread copepods is *Calanus finmarchius*, a vital food for fish, seabirds, and other animals. The seabeds here are rich in oil and gas, which means a lot of seismic surveying. Norwegian researchers found that the effect of seismic air guns on these copepods was much less dramatic. Within 5 meters or so of the air gun, the copepods' mortality was consistent with the Australian results, but beyond that, the little animals seemed mostly fine. And they raised the question whether the Australian results were not from the seismic sounds but the disturbance of the sampling boat. Or perhaps these copepods were more resistant to the air-gun sounds. Ongoing work is now teasing apart exactly what sizes and types of plankton may be injury-prone, and at what frequencies.

When dealing with plankton and invertebrates, another factor to consider is whether an animal can move away from noise. Marine mammals and fish may move away from a sound to escape, but shell-fish, many crustaceans, and plankton have markedly less mobility. These invertebrates are often victims on whom the impacts of noise are least understood.

THE BASS STRAIT separates the state of Victoria, Australia, from the island of Tasmania. More than twenty oil and gas platforms dot the waters, and seismic surveys take place regularly. The strait is also a fishing ground for scallops and lobsters, and in 2011 the Tasmania Scallop Fishermen's Association blamed a recent seismic survey for the loss of 24,000 tons of scallops and tens of millions of dollars in income. Had the air guns killed the scallops?

Jayson Semmens is a marine biologist at the University of Tasmania, and he wondered how seismic air-gun sound might affect commercial species. He was also with IMAS, so he collaborated with McCauley and others to set up an acoustically rigorous test. The team collected wild lobsters and scallops and moved them to test sites in the sea alongside a hydrophone, a geophone, and a video camera. Then they fired an air gun in the water above and gave the animals a thorough exam.

No lobsters died. Many females were full of eggs that developed normally. But the rest of the news was less good.

Basically, the lobsters lost their balance. These crustaceans have a reflex to right themselves if they're flipped on their back, but after the air guns were fired, they took more than twice as long to do so. When Semmens looked closely at the lobsters' statocysts, fluid-filled sacs lined with sensory hair cells at the base of the antennae, the hair cells were damaged—the hairs "sheared off," as he puts it. Since statocysts sense gravity and orientation, this would obviously upset their coordination. Lobsters' blood chemistry changed, too, in ways that suggested a trauma response, leaving them more vulnerable to infection.

As for the scallops, changes in their "blood," a fluid called hemolymph, suggested chronic compromised immunity. And many scallops later died, with deaths peaking three months after the experiment. That pointed at trauma with a long tail, the kind that doesn't kill outright but insidiously. If one didn't know what to look for, a casual glance might have suggested the scallops were fine.

Semmens says he could *see* the air gun had caused a physical shock. The lobsters were on a rocky limestone site, while the scallops were sitting on a sandy bottom, and in the video of the scallop test, Semmens saw the sand jumping with the force. Whether the physical shock or the stress of the sound caused the problems in the animals, he couldn't tell. Either way, there was a literal impact.

Semmens got a lot of "grief," as he puts it, from the oil and gas industry for this study. They claimed a single gun wasn't equivalent to an array and the water was far shallower than real-world conditions. The only way to know was to use a commercial array. So that's what he did next.

Working with a seismic prospecting company is logistically challenging, as McCauley describes. But Semmens was determined. He even included juvenile lobsters in his study, understanding that effects on young animals could have bigger cumulative effects throughout the life cycle. (Adults live fifteen or even twenty years and molt their exoskeletons once or twice each year, but juveniles, little miniatures complete with a hard carapace, must molt much more frequently as they grow.)

Measuring the effect of anything on an animal can be vexing, especially invertebrates. Anything short of death or obvious wounds is difficult to measure, and often goes unnoticed.

Meaningful regulation needs data. Yet to know what a sound does to an animal means measuring a lot more than a sound's "loudness." If it's impulsive, it may jolt the animal more. Its frequency matters depending on hearing range. Even the sea bottom can make a difference to the study: Hard rock could jolt an animal more than soft mud. How old is the animal? Are its gametes like sperm or eggs affected, with implications for its later breeding success?

And as if all that weren't tricky enough, even measuring the sound itself isn't straightforward. "Decibels," which are often used to describe loudness, may be one of the most maddening units of measurement you've ever encountered.

THE KILOGRAM, FROM 1799 until 2019, was defined by "The Big K"—a lump of metal in a secret vault outside Paris. Such standardized units of measurement, across science, bring order and consistency. But decibels are not quite so simple as such other measurements.

"So, decibels," says Jim Miller, an engineer with the University of Rhode Island. "Decibels in the air are not decibels in the water." Humans often blithely compare a sound's loudness to examples like a jet engine or a rock concert. But if it's underwater sound, these examples are not comparable. That's because water and air decibels are calculated on completely different scales. I asked Miller why a unit can vary so much.

Measuring a sound in decibels means measuring the pressure of the wave relative to the reference pressure of the medium through which the wave is traveling. Ambient air pressure is about 1 atmosphere (atm), but in (denser) water, ambient pressure is twenty times that. Ergo, decibels are different in different mediums.

"One way to compare them is to just add sixty," Miller says. Take an 80-decibel noise, what you might get working in a factory. Underwater, that same loudness is measured to be 140 decibels. That's why scientists note whether the measurement is in air or water.

It gets more confusing. Decibels don't increase along a straight steady line, but instead on a logarithmic scale: Loudness increases in multiples, forming a rapidly rising curve. To increase the power of a sound ten-fold, you go up by ten decibels, and to increase it a hundred-fold, you go up by twenty decibels. (It's worth noting that doesn't necessarily translate to how a sound is perceived. Humans, for example, typically perceive a ten-decibel increase as only about a doubling of loudness.) Finally, decibels describe different things. Scientists can measure either the *source* level (the loudness at the source) or *received* level (what the animal actually hears, wherever it is). It's like the difference between measuring the sound of the speakers at a concert from a distance of one meter, and measuring the concert's din at an apartment 5 kilometers away. Miller says scientists also use decibels to describe something called the sound exposure level, or SEL. SEL measures the received energy *and* the duration of the sound. This lets scientists compare sounds of different durations, in terms of total acoustic energy.

Take the challenge of measuring air guns. A pretty standard air-gun noise can be estimated accurately. But *arrays* are trickier. The actual sound level around the array is often some 20 decibels lower than its stated loudness if you calculate it as a point source.

The guns' explosive noise produces a wide range of frequencies, from low wavelengths that may hurt nearby lobsters and scallops to higher frequencies that might be problematic for nearby marine mammals.

Semmens, McCauley, and their team set out in December of 2021 for four days of testing in the Bass Strait. It was a logistical challenge. Semmens and the team had to keep the lobster, octopus, and other test animals in aquariums at the university ahead of time to ensure they were all healthy. Then they had to transport some five hundred animals to a boat in the Strait, with a two-day ride to the survey site, where they didn't know exactly when their experiments would start.

"They were doing a survey beforehand for several months," Semmens recalls. They would wrap up the project when they saw migrating whales arrive in the area, and that was when they could squeeze in Semmens's team's tests. "So we just had to be ready. We had to get the animals in, get them ready, look after them." Once the experiment began, Semmens and the team had four days to do their tests. Each second, each hour, each pass of the ship counted.

They had to have backup plans if the weather turned, if the equipment broke. Each morning, Semmens had a video call with the seismic ship, and then the ship would start running in its tracks, making its great wide turns, and would not be able to stop. Predicting what "sail line" it would traverse next was critical, because the turns were so wide the ship's path was like that of a Zamboni resurfacing an ice rink. As the ship bore down, Semmens had to boat in front of it, lower the cages with the animals to the bottom, and get out of the way. Then, after the ship and its long streamers passed, he had to race back out, haul up the animals, and keep track of everything.

"I'm sure my hair got grayer," Semmens says. But at the end of the experiment, the team had the data.

Their studies have shown that after air-gun exposure, young lobsters molted less than expected, signaling their growth had slowed. They were too tiny for Semmens to examine their statocysts, but he guessed they might be damaged since they took longer to right themselves. Again, they didn't die, and at a glance they looked fine. But they were hurt, with consequences for their long-term survival. His data is still emerging from these field tests.

There have been some technological advances to blunt the noise impacts of air guns. Some can weed out these higher frequencies, or fire in a less impulsive thud. And by firing guns in a staggered pattern, the sound from an array at any one moment may be less. By using vibrations instead of explosions, noise can also be reduced. Many countries, like Canada, have nonetheless put moratoriums on offshore oil and gas exploration. This method of extraction is risky, costly, and in many places, becoming less economical, and while in many areas seismic surveying noise is a concern, globally, it's happening less than in previous decades.

However some marine bioacoustic research is pivoting to another impulsive sound, ironically, coming from a structure intended to replace oil and gas.

I THOUGHT I knew the scope and size of wind farms. Growing up in southern Ontario, I'd seen their slender white blades begin to poke up here and there above the pastoral hills near Toronto in the early aughts. In the windy waters off Kingston, low, grassy islands in Lake Ontario sprouted forests of turbines as the province stumbled toward green energy.

But as I flew into Amsterdam Schiphol in 2022, I understood the Europeans are masters of the offshore wind turbine. The Princess Amalia Wind Farm sprouted in the sea beneath the flight path—huge turbines arranged in neat rows that seemed to stretch up until they nearly brushed the plane's belly. I could see each spin of the huge blades. Farther north, on the shores of Scotland, the casings sitting onshore waiting to be installed were bigger in diameter than many houses.

Offshore wind turbines are *big*. Land turbines can reach up to 90 meters from base to hub, but offshore turbines easily top 100 meters and are getting bigger. In 2022 more than 90 percent of new wind turbines built around the world were built on land. But a large amount of untapped wind-power potential exists in coastal offshore areas.

The first offshore wind farm, Vindeby, in Denmark, entered service in 1991 with eleven turbines. It was decommissioned after twenty-five years, but by then wind had taken off in Europe, with Sweden, Germany, and the UK all constructing farms. From 2000-01, Denmark built the twenty-turbine Middelgrunden farm, and in 2002, Horns Reef, with four times the turbines of its predecessor.

In 2021, the U.S.' Biden administration injected $3 billion into offshore wind, but there was already a rush to survey the sea for suitable sites. The first offshore wind farm in the United States was built in 2016: the Block Island Wind Farm, about 4.5 kilometers southeast of Block Island, a small island between the tip of Long Island and Martha's Vineyard. In 2020, a pilot wind farm began operating off the coast of Virginia, with plans to build a 176-turbine farm by 2026. The U. S. Bureau of Ocean Energy Management (BOEM) administers leases on blocks and tracts of seabed, wherein suitable sites are surveyed, and leases are granted to companies or consortiums thereof. Many leases have been granted, and construction is under way on more ambitious farms, off the coast of the northeastern U.S.

Concerns about turbine noise and its impact on humans arose when farms began proliferating on land. They made a constant low thrum that some people found so annoying they sued. Offshore farms make so-called operational noise too, largely from the machinery in the nacelle that travels down through the tower, the foundation, and out into the water and seabed. This noise isn't impulsive, more of a continuous drone, and there isn't a lot of evidence that it's actively harmful to many animals. For better or for worse, some creatures are even attracted to a sound that may mean home or food. Offshore wind piles, the base of the turbine that stands in the ocean, often get a lot of life settling on them, just as ship hulls are magnets for biofilms, barnacles, and other small animals. (Shipping firms spend millions on anti-fouling hull paint and hull cleaning to rid their fleets of fuel-efficiency-killing animals that seem to think the ship is a giant metal home reef.)

But the noisiest aspect of wind farms is their construction.

There's been research into, and even a few prototypes of, turbines that rest on massive floats tethered to the seabed. These designs are meant for deeper water, where a pile isn't feasible. But the great majority of wind turbines under operation—or planned for the future—stand on massive piles that must be driven into the seabed by strikes of a hammer.

A driven pile makes an impulsive bang, just like an air gun. This noise spreads into the water, with all the attendant issues. But it also radiates out through the seabed. And one new bioacoustics frontier is the blurry space between sound and vibration—and all the fish and invertebrates on or above the seabed who don't sense sound pressure but particle *motion.*

IN SEPTEMBER OF 2015, 4.5 kilometers southeast of Block Island, the ocean is 30 meters deep, about the height of a ten-story building (or the length of a blue whale). Five turbines are being installed here on this underwater bank. Each turbine will generate 6 megawatts of power, and so the whole thing will be 30 megawatts. It's not a huge wind farm by European standards, but it's the first offshore farm in the United States.

These turbines will rest on structures called jacket piles. If the water were very shallow, they might sit on a gravity base—a giant concrete block that rests on the bottom. A bit deeper and they would likely be monopiles, which, as the name suggests, are just single piles, between 3 and 8 meters in diameter, driven straight down 40 meters into the muck. But where the water is deep and turbulent, or the makeup of the seabed less stable, a jacket is used—a sort of latticed metal pylon with four legs. The supporting piles will thus be driven not straight down but at a 13-degree angle.

It will take between five hundred to more than five thousand strikes to sink these meter-plus-diameter piles, up to depths of 75 meters. A pile driver is basically a giant hammer, hitting between fifteen and sixty times per minute, using brute force to hammer the base into the seabed.

The federal government owns the seabed and leases the tracts to companies to build the wind farms. The agency responsible is BOEM, which has an acoustic monitoring program at the Block Island site called Real-time Opportunity for Development of Environmental Observations, or RODEO (acronyms strike again), under way. RODEO is currently listening with a towed array of hydrophones a kilometer or so from the construction site: a small sled on the bottom of the sea, half a kilometer away measuring sound and vibration, and two hydrophones, 7.5 and 15 kilometers away respectively. These instruments, along with limited studies of certain species, provide a glimpse of how a sound wave might spread as piles are driven in these waters—and who it might encounter.

The piles are driven into the seabed from August to October, four piles per turbine, twenty in all. Each jacket pile requires about three times more strikes than a monopile. The hammer strikes a pile approximately 1.5 meters in diameter, a hollow tube with a wall as thick as a bagel, with enough force to slightly deform it, a bulge starting at the top and traveling down its length. This bulge presses against the water, oscillating as the pile rings with the force, and sends a sound wave out into the water, starting at the top and moving down, an upside-down cone of noise spreading outward. Each strike, at least close to the source, makes an impulsive sound.

This sound attenuates more quickly in the summer as it spreads out. The surface is warmer than the depths, which sends the sound refracting downward. In winter, the water is well mixed and more uniform; the sound spreads straight out. But in early autumn, it goes a medium distance. It's a good 20 kilometers before the background noise of the windy, wavy, current-churned autumn Atlantic drowns out the thuds.

In the immediate vicinity of the pile, maybe some plankton are knocked out by the strong particle motion of the water. Maybe some larval animals are stunned. But farther away, the sound has attenuated somewhat. If it meets a fish—say, a salmon or a cod—the fish can hear it, but likely won't be wounded. Only if the cod is close enough to receive the sound at 207 decibels is it likely to cause injury. The salmon likely won't be injured until 213 decibels.

As the noise heads farther out into the sea, perhaps it encounters a longfin squid, shaking its body relative to its statolith, well within its sound-detecting abilities. If this happens within a few kilometers, the squid might squirt a jet of ink or move away, but if the noise continues more than twelve or thirteen seconds, the squid gets accustomed to it and stops squirting and jetting. This reduces the squid's energy output, but if an actual predator comes along, the squid may ignore that sound, too.

The pile driving noise moves farther, becoming less impulsive with distance. Its thuds are well-audible to a male fin whale, passing through the ocean beyond the last buoy.

As the sound spreads in the water, something else happens in the seabed. The pile sends seismic waves down and out. The first, and the fastest, is a p-wave, a sound wave of particles move back and forth, compressing the medium of the seafloor. The second is a transverse s-wave, where the particles move up and down. These surface-interference waves travel along the seafloor and interact with the water above them, forming evanescent sound waves that stay within about a meter of the seafloor.

In this complex environment, sound using pressure measured above the seabed won't simply correlate to the motion of the particles. Yet the invertebrate animals that live on and in the seabed, and those in the water column, are likely sensitive to this particle motion. This means when we measure sound in such places or circumstances with just a pressure-sensing hydrophone, which is most hydrophones, we are measuring a stimuli irrelevant to these animals.

The sound and its substrate vibrations reaches a scallop, which "coughs," opening and shutting its valve. It reaches a hermit crab that flicks its antennae and skitters away. A plaice, a flatfish like halibut or flounder, lies flat on the seabed, blending in with the gray-brown. In addition to its ears, its sensitive body detects vibrations from the seabed, too.

The life cycle of entire animal phyla takes place between grains of sand on beaches or on seabeds. These loriciferans, nematodes, flatworms, kinorhyncha, water bears, and rotifers live in the fluid-filled space between grains. As the Australian work shows, even very small creatures can be impacted by noise. What do these meiofauna detect? How do these seabed vibrations affect them?

Just as animals underwater tap into sound because it's the informational channel of choice, so animals on the seabed may tap into substrate vibrations. Why wouldn't they? Vibration is known as one of the oldest communication systems in the animal kingdom. Insects use it widely, sending shivers up and down leaves and plant stems. Mammals likely detected ground-borne vibrations in their ground-level jaws before they developed in-air hearing. We are just beginning to measure and study vibrational communication, and on the seabed the line between sound and vibration can be blurry. The anatomical structures that can detect low-frequency sound can almost certainly feel the shudders of earthquakes, fin-whale song, and driven piles. One of the biggest data gaps in the sea is how sound and noise affects fish and invertebrates that detect particle motion, and scientists are now measuring particle motion and seabed vibration in different conditions, working to reveal what these animals detect.

Ultimately, the noises most likely to cause big problems for all animals are the high-energy shocks: the blink-punch of seismic air guns, the unpredictable shriek of naval sonar, and the rhythmic, relentless pounding of piles for marine construction of wind farms, docks, and other structures. However, these have the dubious benefit of being somewhat restricted in space and time.

But there is another problematic sound under the water, the source of a very great deal of noise: ships and boats. Our vessels are everywhere, almost never quiet enough to be forgotten.

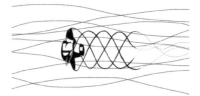

CHAPTER 9

Shipping: The Global Growl

"All the while the world is turning to noise."

—Peter Gabriel, *Signal to Noise*

LATE SEPTEMBER, AND AUTUMN COLORS flush the forested slopes of Quebec's Laurentian Mountains. Sun dapples the two-lane blacktop that winds above the brilliant blue St. Lawrence River. The St. Lawrence is a marine highway that links Toronto, Cleveland, Chicago, Thunder Bay, and other Great Lakes ports to the Atlantic. Here run cargo ships, vehicle carriers, and cruise ships, alongside recreational boats.

I've followed the river northeast past the neon riot of Montreal and the historic grandeur of Quebec City, to Tadoussac. Here, the river widens, shading into the Gulf and the salty Atlantic beyond, just where the smaller Saguenay River meets the St. Lawrence's northern bank. Fjord, river, and sea meet in a restive churn that stirs up a rich food web. In summer, whales from thirteen species mingle below. Minke whales, blue and fin whales, humpbacks, and dolphins ply the waters.

At the Pointe-Noire headland, a network of viewing walkways zig-zags through maples, evergreens, and golden birch. Tadoussac offers scientists fruitful research sites alongside impressive whale-watching, and the wooden placard at the walkway entrance makes it clear that belugas are the local celebrities.

Most belugas live in icy Arctic waters. As the last ice age ended, a handful lingered in the St. Lawrence Estuary instead of following the cold water north. Now their descendants are the world's southernmost, cut off from the rest of their kind. Quebec fishermen long hunted them for sport, their white hides fetching $15 each. Today they number around nine hundred.

A beluga surfaces below the platform, then dives into the sapphire water. Its white form shades to teal to royal blue as it vanishes in the depths.

As our local guide speaks, I move closer to hear her clearly, in part because of the wind, in part because a cruise ship is approaching from the St. Lawrence.

All afternoon we've watched vehicle carriers, cargo ships laden with ore and grain, and container ships moving cars, iPads, and other consumer goods as they pass. The Saguenay also draws tourists, particularly during fall, so cruise ships abound.

So, the nearby federal marine park limits how close commercial and recreational vessels can come to whales. Ships must stay 400 meters away. If whales come closer, boats must slow until the whale leaves. Whale-watching boats can get a special license to creep within 100 meters, or 200 meters of belugas or blue whales, the two endangered locals.

I think back to high school track-and-field and the 100-meter dash, my clearest visualization of this distance. Four times that is pretty far. This all sounds perfectly reasonable.

Our guide has to pause. The cruise ship is too loud. It's four stories of engine noise. It's so big that as we stand on the slope we look across at its decks, not down, as the beluga dives. Up, down. The ship passes up the fjord, taking its noise slowly upriver.

The Saguenay's mouth is a kilometer wide. The beluga was swimming just off a flanking headland. Therefore, the ship passed 500 meters from the beluga. Perfectly legal, with room to spare. But . . . is that *really* what 500 meters looks like? If the beluga was under for a few minutes, would the ship have seen it, let alone stopped? How could *anyone* actually track the belugas underwater?

Previously I had trouble linking the concepts of sound and physical harm. I can picture other threats like ship strike or oil spills. But this is the moment I first realize the acoustic power of noise. The beluga has been down a while. Is it scared? Stunned? I see its white form surface, breathe. It keeps diving. Up, down.

Unless you live near a port or work in the industry, you may not give maritime ships much thought. Shipping is an opaque industry. The vessels that carry 80 or 90 percent of global trade do so far out in the ocean, or near restricted ports and docks. In Vancouver, orange gantry cranes at the container docks of Vanterm and Centerm rear their giant heads above downtown, and freighters moor in English Bay like a herd of great iron cattle, but that's as close as most people get. Victoria lacks container docks, but we do face the Strait of Juan de Fuca, a major lane for every ship bound for the ports of Vancouver, Seattle, or Tacoma. Often, four or five ships are in view, distant and small-seeming as toys.

The Ogden Point seawall juts out into the strait. The outer side of its 800-meter length falls off into a fringing bull-kelp forest. The inner side sees multiple cruise ships each day during the summer months, each the size of a huge apartment complex.

In 1980 the world's merchant shipping fleet (meaning *all* ships, not just containers) numbered just under 700,000. In 2020, it's more than 2 million.

Globally, container ships, with their colorful, LEGO-like boxes, make up only about 10 percent of commercial ships. The rest are vehicle carriers, cruise ships, oil tankers, or bulk carriers toting ore, coal, grain, or other commodities. Container ships nonetheless carry about half of the world's cargo, thanks to economies of scale. Containerized cargo moved the global shipping industry to efficient, standardized "Twenty-foot Equivalent Units" (TEU), or a twenty-foot-long corrugated metal container. These containers let us ship more goods around the world than ever before. We all rely on them, yet most people rarely see these containers and ships.

In January 2022 the container ship SM *Busan* (6,622-TEU capacity) docked at Ogden Point in Victoria for emergency repairs after losing power in the Pacific. I leaped at the rare chance to see a giant vessel up close. Its great stern towered above the dock, so high that falling from its back railing would be deadly. Walls of blue and red and green containers were stacked six high and sixteen across—more than double the height of the stern. *Busan* was 304 meters long (ten large blue whales). And yet as container ships go, she isn't large. Hundreds carry more than 10,000 TEU. Eighty have more than 20,000 TEU capacity and add another hundred meters to SM *Busan*'s length.

Between 1992 and 2012 global shipping traffic increased fourfold. The Panama and Suez Canals cut weeks off lengthy trips. Ships that can't fit in these canals trundle across the Pacific, from Long Beach and Los Angeles, Seattle and Tacoma, to Busan and Shanghai, in lanes like the Strait of Juan de Fuca, as technologies like AIS and GPS track them 24/7.

Of course, they make noise. Average ship sound levels have doubled in the ocean every decade from the 1960s through the late 2000s.

Standing on the deck of a ship, the loudest sound you hear might be the growl of the engines. Yet beneath the water, ship noise comes primarily from the propeller. Specifically, from bubbles that form on the prop's edge in a process called cavitation.

As a propeller spins, the blade's leading edge presses the water, creating high pressure, and the trailing edge of the blade follows with very low pressure. We know that water boils at lower temperatures as ambient pressure decreases. (This is the same principle behind our snapping shrimp's bubbles. Snapping its claw shoots out a jet of water that trails extremely low pressure in its wake, flash-boiling the seawater. The bubble collapses, and pops loudly.) On a ship's propeller, pressure drops at the trailing edge, and bubbles form. They collapse on the blade or trail off into the water. At the same time hydrodynamics at the root of the blades makes the prop hub trail another line of bubbles called the hub vortex.

Invisible below the water, *Busan*'s propeller is several meters across. The largest container-ship props can be 9 meters. These propellers shed bubbles that flash-boil seawater and oscillate, the vibration becoming a sound. Bubbles vary in size, so collectively they make many frequencies, most in the low hundreds of hertz. Ship noise, then, is broadband, with most of the energy in lower tones.

Ship sound isn't necessarily deafening. It's more of a chronic background drone, like living near a highway. In the bull's-eye framework of zones of influence, ships often fall in the outer rings: influencing the animal's behavior or communication, making an animal move to where it wasn't going to go; or interrupting it echolocating, singing, or vocalizing.

Think back to the zones of influence. Most animals are far enough from a ship that its noise will merely reduce the animals' listening and communication space. Listening space is the space within which an animal can hear signals. This is often thought of as a sphere, but if the sound is directional, the shape of the listening space can change depending on whether the animal is facing it. Echolocation clicks might have a different space than other vocalizations.

Communication space, conversely, is the distance over which the animal can receive and decipher signals from others. Listening and communication space can shrink when a noise falls in an animal's hearing range. For mammals, in the cochlea, remember, the membrane is vibrating at a particular spot, a specific frequency. If another sound occurs at that exact frequency and those adjacent to it, the target sound is masked: The animal can't pick out the original signal.

An earthquake, or a hurricane, can do this. A shipping lane, like a highway, also shrinks this sphere. As for invertebrate animals that can't move away, scallops or mussels or seagrass or polychaete worms that live beneath marine highways, chronic, long-term noise can mimic short, intense noise. One of Semmens's more alarming findings occurred when the team collected test lobsters from beneath a shipping lane. Their statocyst hair cells' cilia had already been sheared off. Loud ship noise can be like working in a factory for decades.

Faced with chronic ship noise, the animal responds in one of a few ways. In some cases, mammals can compensate and increase their hearing sensitivity to certain sounds. The animal can also move somewhere quieter. Or it can call in a different frequency, call louder, repeat itself, or stop calling altogether. These strategies come with costs. Sometimes the costs are negligible. Sometimes not.

Take, for example, the cleaner wrasse. It's a finger-sized fish with brilliant black-and-blue striping, and one of the most enterprising fish in the sea.

Kieran Cox's current supervisor, Isabelle Côte, did a study on wrasse in the Pacific Islands of Moorea. These fish have a symbiotic relationship with larger fish that would normally eat them. The wrasse set up "stations" on the reef to which other fish come to get cleaned. The wrasse eat parasites and dead scales, keeping the client healthy. Moorea also hosts fishing and reef tourism, and motorboats often pass over this bustling marketplace. When Côte watched the fish during test runs of boats, the noise seemed to distract the wrasse. They didn't stop working, but they cheated, stealing more healthy, juicy scales from the client. Wrasse seem to do this when they can get away with it. Generally, the client chases the wrasse, teaching it a lesson: A chastened wrasse cheats less for a period of time. The wrasse cheated more around boat noise, perhaps because the clients were distracted. The noise wrought a subtle economic change.

THE NEWSPAPER CARTOON *Family Circus*, the most widely syndicated panel in the world, features a nuclear family with four kids in a single frame. It sometimes runs a "dotted line" strip, where a line shows the tracks of a character meandering throughout the house, yard, or other setting. The D-tag data from killer whales, diving for salmon in the Salish Sea, reminds me of this cartoon—a twisting, three-dimensional line tracing the whale's path through space.

A D-tag is about the size of a small paperback, with suction cups that attach to a whale's smooth skin. Triggered remotely, the cups release and the tag floats to the surface, where boats zero in on its GPS signal. A skilled person standing in a boat holding a long pole with the tag on the end taps the tag on a whale when it surfaces. The tag travels with the whale for a while. Then, once retrieved, the D-tag provides audio, GPS location, and acceleration: the story of a whale's movement, the sounds it made, and the sounds it heard.

Killer whale salmon-hunting dives have a unique profile. The whales swim near the surface, making slow clicks, then dive down in the depths corkscrewing and twisting, chasing the fish, their clicks shading into those terminal buzzes that precede capture. When they approach their prey, they suddenly accelerate, or jerk, and roll to one side. When they chomp down, debris often floats to the surface as the satiated whale zips back upward to find its family and share its catch.

Thanks to ever-pinging automatic identification system (AIS) data, the location and speed of commercial vessels is largely public, including whale-watching boats, recreational boats, and other vessels around the Gulf Islands, where the Southern Residents come to eat salmon through the summer and fall. The AIS data and D-tag data together tell a tale.

When vessels were closer than a kilometer and a half, the whales dove faster but caught less prey. This wasn't related to the vessel's sound. The boat's presence was enough. One possible reason? Most small vessels like Libby have echo sounders that ping, above human hearing but right at the peak frequency of a killer whale's echolocation click. When vessels used their echo sounders, nearby whales dove slower and stayed down longer.

When there were noisy vessels nearby, the whales made more of the slow and searching surface clicks. The females in particular made fewer buzzes, suggesting they didn't make as many chases and captures. The males, however, dove for longer. Preliminary estimates show that for every decibel of increased noise underwater, the whales buzzed 21 percent less. The boats are well within local regulation, but for an endangered animal that eats one type of food, noise has caloric consequences. Simply put, whales are expending more energy and eating less around boats.

A similar drama plays out in the waters off Denmark, where the harbor porpoise echolocates for fish amid the busy shipping lanes feeding the North Sea, Baltic Sea, and Kattegat. Suction-cup tags tracked seven porpoises foraging for four days around ships, and how their behavior changed. The porpoises spent about 59 percent of the time foraging, and the porpoise in the quietest waters heard ships rumbling 17 percent of the time. The loudest porpoise was bombarded by noise a whopping 83 percent of the time. When the noise edged over about 100 decibels, the animals buzzed less often. During one pass of a ferry, a porpoise stopped foraging while the ferry was still 7 kilometers away, waiting fifteen minutes before it began to hunt again.

In many Western countries today, for many privileged people, calories are framed as a surplus to be managed, and skipping a meal isn't a concern. Yet porpoises don't exist like this. If they want to eat, they need to hunt, which takes time and costs calories.

Maybe the porpoises are adapted to these busy waters. But maybe they have no choice but to try to hunt, despite the noise, if their prey congregates in hot spots. Porpoises' metabolic needs are high. If they are interrupted twelve times a day for six minutes, that's a real energy cost.

Some populations of harbor porpoise are thriving, but others are not. In the inner Baltic Sea swim some 500 animals, an isolated and dwindling population. An already stressed or sick animal cannot afford to lose time, and calories, to ships.

IF SOUND EXPANDS the world, then noise shrinks it and takes it away. What's an animal to do in noisy waters when it needs not only to listen but to communicate over distance?

Valeria Vergara came to Tadoussac in the summer of 2017 with the same tower setup that she'd used in Cumberland Sound. But this time she was watching a calving ground—safe and sheltered waters where mothers come to give birth and nurse their newborns. Twenty-two kilometers up the river from the mouth of the deep, narrow Saguenay, the Sainte-Marguerite River creates a shallow delta where mothers hang out with their calves, eating, resting, chilling—and calling.

Vergara and her colleagues had estimated the sound levels of rudimentary calls made by a newborn beluga—120 decibels, about 22 decibels quieter than adults or subadults. If this data, gathered from a newborn and adults at Oceanographic Aquarium in Spain and four wild belugas in the St. Lawrence, translated to all belugas, then other belugas potentially heard their faint newborn calls up to only about 350 meters away, while those of adults had a communication range of up to 6.5 kilometers.

Vergara spent twenty-three days on the tower, counting boats and belugas and recording noise levels with a hydrophone, and found boat noise halved the contact-call range. A baby whale bleating into boat traffic could be heard over less than 200 meters, and an adult less than 3 kilometers. And that was only for *hearing* the call. Picking out the call type, its details and nuances? For animals whose vocalizations may include dialects and identification calls, animals whose calls have a range of frequencies, probably important information, even identification information, the reduction in communication space could be even greater, since another beluga not only needs to hear the call, but also to understand it.

Then there was the effort of shouting back. In the early 1900s, a French doctor named Étienne Lombard identified an effect that bears his name. The Lombard effect happens in humans, birds, and marine mammals when there's background noise. We vocalize louder to be heard over the din. In the early 2000s, biology student Peter Scheifele found that the St. Lawrence belugas did this around boat noise. Vergara wrote up her observations in a paper titled, "Can You Hear Me?"

There seems to be an obvious fix to ship noise as a pollutant: just be quieter. That's a seductively easy-sounding idea. Of all the ways to hush a ship, there are three big ones: (1) physically separate ships and animals by shifting shipping lanes and setting up warning systems, (2) slow down, or (3) design a quieter ship. Simple, right? As a project done by shipping giant Maersk demonstrates, the latter is not always straightforward when it comes to the biggest boats on the planet.

Globally, the Danish firm A. P. Moller-Maersk operates some seven hundred vessels. They were the world's largest container-ship company from 1996 until 2022, when they were surpassed by MSC. In the mid-2000s, Maersk formed a working group with some other carriers and their clients to think about how to increase their efficiency. The cost of fuel was high, and some of Maersk's customers were global brands like Nordstrom and IKEA that were growing worried about public backlash for their shipping carbon footprint. For slightly different reasons, Maersk and their clients both wanted efficiency. They studied how to optimize their networks, as container ships run on set schedules and routes, and they ordered new vessels with efficiency upgrades.

A container ship's life is twenty to twenty-five years. If they waited for new builds to incorporate designs, the fleet wouldn't be fully upgraded for decades. So, Maersk undertook a project they dubbed the "radical retrofit," which would give some of their huge container ships a makeover. "We decided to retrofit certain of our vessels that were, sort of, teenagers," says Lee Kindberg, North American head of environment and sustainability at Maersk. "They still had lots of life in them." Every five years, a ship gets dry-docked, repainted, cleaned, and given upgrades or repairs to keep it in working order. Why not make them more efficient while they were at it?

Maersk retrofitted 100 of their ships over the course of five years. Twelve of these were "G-class" ships (all with Danish names beginning with G—Gunhilde, Gertrude, Georg). For example, Gudrun (built in 2005, retrofitted in 2015) will likely operate for another ten years. "The last one—which I think was Gunhilde—was done about three years after the first one," Kindberg says. Each G-class vessel is 367 meters long, 43 meters wide, with a draft of almost 16 meters. "Yeah," Kindberg says fondly. "They're big girls."

Each vessel is also an immensely complicated system of moving parts, and changes to one thing—say, the propeller—affects trim, which has knock-on effects on efficiency. Bow and hull shape affect wash along the hull. Considering that dry-docking something the size of an apartment building is very expensive and cuts into the ship's working life, only a comprehensive suite of carefully integrated changes is remotely cost-effective.

At the outset, noise considerations weren't part of the retrofit. But at the time, Kindberg was starting to work with a unique initiative in

Vancouver called the ECHO program, which tried to reduce shipping noise in the Southern Resident killer whale territory by asking ships to voluntarily slow down in certain areas of the Salish Sea. Because of this, Kindberg wondered if the more-efficient ships might also be *quieter* post-retrofit.

For ships, as for any other machines, sound and noise is lost energy. "Anytime you have noise happening, it means you're not being as efficient as you can be," says Kindberg. It made sense that a more efficient ship would be quieter. Specifically, the retrofit had outfitted the "girls" with more efficient propellers, traditionally the loudest part of the ship. These new propellers reduced noisy cavitation.

For a large ship, one of the most common propeller upgrades is a boss cap fin. It looks like a tiny second propeller, mounted on the hub of the main prop, and it reduces the hub vortex to almost nothing. Designers can also change the angle or the number of blades, improving efficiency and reducing cavitation. Shipbuilders like to avoid cavitation, regardless of noise. Each bubble that forms along the prop can pit the steel and reduce its efficiency.

To discover if the new props were quieter, Kindberg needed to listen to her spruced-up fleet in action. "You'd think that [noise] would be something that would be routinely measured during sea trials," she says, "but there really aren't a lot of facilities around the world where you can go and have it measured in a standardized way."

In fact, measuring noise from a container ship, just like retrofitting itself, is a surprisingly complex undertaking. Should it be measured from the front, where it's quiet? The back, where it's loud? Directly beneath? To the side? You can't just put a hydrophone in the water and hit Record. So, the noise-reduction question remained open until a fortuitous conversation. "A woman at the Vancouver Aquarium," Kindberg recalls, "mentioned to me that John Hildebrand at Scripps had a lot of data on underwater noise."

Hildebrand's lab monitors a suite of underwater hydrophones called HARPs off California, including one lying in 580 meters of water beside the outbound shipping lane from the Port of Los Angeles and the Port of Long Beach. These two ports sit side by side just south of L.A. and are the biggest container ports on the continent, the North American gateway to Asian consumer goods like clothing, couches, and iPhones. This makes the Santa Barbara Channel a superhighway, where ships start their two-and-a-half-week jaunt across the Pacific.

Kindberg knew her G-class vessels were "workhorses" on the Pacific routes. (Their retrofits included shore power, a technology that lets ships turn off their engines at a dock and plug into a shore-based power source, reducing emissions near coastal areas. It's cost-effective to keep a ship on routes that have this option, which many Pacific ports do.) Gudrun, Gertrude, and their sisters spent their days crisscrossing the Pacific. "And I realize that at least one of these vessels calls at L.A.— Long Beach. And [Hildebrand] had, like, ten years of data." So Kindberg asked to combine his recordings with the AIS data pinpointing the passage of the G-class ships, and then compare their sounds before and after the retrofit.

Many things changed in the retrofit, so they measured two frequency bands—from 8 Hz to 100 Hz, and from 100 Hz to 1,000 Hz. Lower-frequency changes come mostly from propeller changes. Higher-frequency changes came mostly from the engine and machinery tweaks.

The preliminary data showed that post-retrofit, the source levels of container ships were six decibels quieter in the low-frequency band. The data was only for five ships, and it wasn't peer-reviewed, but it was the only real-world data of its kind. Then a few years later, Vanessa ZoBell, a PhD student of Hildebrand's, crunched the numbers for the other G-class transits.

The source levels of the twelve ships were five decibels quieter post-retrofit, in the low-frequency band. Yet the radiated noise, the noise measured at the hydrophone, was *louder* post-retrofit, by almost a full decibel. As with everything else in naval architecture, it's complicated.

ZoBell explains that the ships were now carrying an additional 1,100 containers each, sinking them deeper in the water and increasing their draft. This increased draft then changed how propeller sound moved outward, thanks to something called a "Lloyd's Mirror" effect, caused by sound interference at the sea surface.

"So basically," she says, "since the source is close to the surface, [some sound] rays will bounce off the sea surface and reflect downward." This reflection then destructively interferes with other sound waves, and they cancel each other out. It's a bit like noise canceling headphones, whose built-in microphones create opposing sound waves to create silence. After correcting for this, thanks to complicated math, there was less noise at the sound source, but more radiated noise overall.

As if that wasn't complex enough, sound spreads out from ships completely differently in deep water (about 150 meters or more) versus shallow water. It's such a critical difference that the International Organization for Standardization has one set of standards for measuring sound in deep water, but is drafting different standards for doing so in "shallow" water.

Ship noise is a mutable thing, depending on the ship's condition, its propeller, its engine, its draft, its tonnage, the reflection of the sound off the surface, and whether it's in deep or shallow water. So tweaking ships to make them quieter is not as simple as it sounds.

When dealing with economies of scale, cargo can get quieter and greener per metric ton, per TEU, per unit. But with more cargo overall, the noise will still increase. Kindberg says that Maersk has reduced greenhouse gasses per metric ton shipped by more than 50 percent, and if ZoBell's numbers are indicative, retrofitted ships actually create fewer decibels per TEU. But the increasing size of the ship and volume of goods make up that difference.

There is another level of noise reduction here: Larger ships make fewer trips. But what this means for animals isn't clear. "So now they're carrying more, so potentially, less transits required," ZoBell says. "But what's important to a whale? Less transits? Or more transits, but quieter?"

SO WHAT ABOUT another strategy: slowing down?

Haro Strait, where naval sonar sent J Pod scattering that afternoon two decades ago, is more than a whale-watching hot spot. It is also a dog-leg shipping route that sees around fourteen ships pass through each day: on a busy day, more than twenty. Haro Strait's island- and shoal-studded convolutions are hazardous enough that all ships require a professional pilot trained in navigating it.

Before any ships enter Haro, they pause on the Magdalen Banks south of Victoria. Then, from the dock behind Ogden Point, bright-yellow boats zoom out to deliver or pick up pilots, who aid the ship's navigation between there and Vancouver. The job is so difficult that pilots train for decades. If you stand on the seawall and squint, sometimes you can see the pilots climbing from the yellow boats up onto the great ships on a ladder slung over the side.

Ships and whales have long had an uneasy coexistence in these waters. Concerns about cetaceans, particularly the endangered southern residents, rose sharply in recent years as contentious proposed pipeline and port expansions promised even more vessel traffic. In 2014, the Port of Vancouver, the Vancouver Fraser Port Authority, led an initiative dubbed the Enhancing Cetacean Habitat and Observation, or ECHO, program to try one of the only ways to make ships quieter: by slowing them down. By 2017 ECHO was ready for a trial.

From August 7 to October 6 that year, pilots heading out to the commercial ships gave them the option of slowing down to a target speed of 11 knots as they set out through the Haro Strait. Several hundred ships, 61 percent of commercial vessels that passed through the Strait participated. Of those that did, 44 percent of all commercial vessel transits managed to get under 12 knots—55 percent under 13.

For a bulker or cargo ship, that was about two knots slower than normal, and about 5.9 decibels quieter at the source level. For a large container vessel that was up to 7.7 knots slower than normal, and reduced their sound source levels by around 11.5 decibels. As for the received sound levels, or what an animal might hear, hydrophones near Haro Strait detected a reduction of 1.2 dB during the slowdown, or about 24 percent reduction in sound intensity (remember, decibels are logarithmic).

Slowing down has some costs. As Kindberg described, container ships run on schedule, like a bus route. Delays at sea or in port make carriers noncompetitive. Then there's safety. Sometimes it's dangerous to slow down in high currents or bad weather. Finally, the slower a ship goes, the longer it's in an area. That might translate to more cumulative noise, more greenhouse gases, and other toxic emissions near shore. If a ship is in poor maintenance, then staying longer in the area won't be quieter and could make things worse, though by listening in real time, such ships can be identified and their owners notified. On balance, a slower ship is quieter, and one study estimates suggest the 2017 slowdown trial data speeds could have reduced the amount of time killer whale feeding was affected by the noise, by about 10 percent. Participation rates in recent years have been over 80 percent, and the program has expanded its slowdown areas and even shifted the shipping lane along a stretch of the route into Haro Strait.

Kindberg likes the ECHO program. It's an opportunity to make her fleet a little less intrusive to the ocean, and the connectedness of modern ships means that she gets up-to-the-minute reports from each ship on speed, fuel efficiency, position, and other metrics when ECHO reports how well "her" Maersk ships did in the Haro Strait. It's like a videogame, with Kindberg aiming for a "perfect" month of speed reductions.

Slowing down also reduces the chance that ships will strike or kill whales. In the Gulf of St. Lawrence, where endangered North Atlantic right whales come to feed, the Canadian government has a mandatory slowdown when whales are in the area.

ONE OTHER NOISE-REDUCTION strategy is to physically separate ships from animals. This has happened in a few places. The Kattegat Strait in Denmark is a shipping lane that sees some eighty thousand ships each year from the North Sea into the Baltic ports. In 2020, Sweden and Denmark decided to shift the shipping lane. Scientists took advantage of this and measured noise on the old and new lanes. The Kattegat is home to the Baltic harbor porpoise, a population of which is endangered, and there were concerns about the impacts of noise on the porpoise's echolocation and feeding abilities. These small odontocetes had been shown to stop hunting when ship noise was too loud, which for a small cetacean that hunts a large proportion of the time, meant they may not be able to get enough food. When the numbers were crunched, the waters around the old lanes were found to be slightly quieter as traffic dwindled. But the new lanes were louder, and the problem had only shifted. It's worth noting that this natural experiment wasn't designed for the porpoises, however. Where shipping lanes run through an animals' hunting or breeding grounds, the noise reduction in those areas from moving the lane might make a significant difference.

On a smaller scale, many places have regulations about how close boats can get to whales. I went whale-watching on the Saguenay one bright fall day. It was my first glimpse of baleen whales, and I was thrilled to see the flukes of a humpback sliding gracefully out of the water as it dove. Smaller minkes popped up everywhere. But privately, I was most excited at the sleek back of a fin whale. I'd grown deeply affectionate toward these second-biggest cetaceans with their monotonous seismic song.

The Saguenay-St. Lawrence Marine Park regulates how close ships and recreational boats can come to whales. From my perspective on the water, the 200 meters that our whale-watching boats were allowed seemed, if not completely unobtrusive, at least polite.

But our little twenty-foot-long open motorboat was soon dwarfed by a three-story vessel with a large deck, and a voice blared through a speaker loud enough that I heard every word. Most of this noise reflected off the surface, sure. But the boat shadowed us for much of the trip.

Our noisy boats have, curiously, isolated us from whales in a strange way. Engine noise drowns out the sounds fishers, boaters, or other ocean travelers might otherwise hear from whales above the waves or through the hull. In the paper accompanying Schevill and Lawrence's St. Lawrence recordings they note, "Arctic voyagers . . . have heard some of these underwater calls under favorable conditions of quiet as the animals swam under their vessels; sometimes the hull, acting as a resonator, amplified these sounds, especially in the past, before engines and generators made even idle ships noisy."

The perspective was echoed by Roger Payne and Scott McVay in their 1971 paper: "During the quiet age of sail, under conditions of exceptional calm and proximity, whalers were occasionally able to hear the sounds of whales transmitted faintly through a wooden hull. In this noisy century, the widespread use of propeller-driven ships and continuously running shipboard generators has made this a rare occurrence."

As we discover more about the sounds animals make underwater and the impact we have on them, the impetus grows to find a solution to noise. Slowdowns, lane shifts, retrofits, and research on a wider range of species are all good starts. But shutting up is not easy. Can humans really learn to be considerate marine neighbors in the coming, critical years?

CHAPTER 10

From Science to Art:
How to Quiet an Ocean

We have lingered in the chambers of the sea
By sea-girls wreathed with seaweed red and brown
Till human voices wake us, and we drown.

—T. S. Eliot, "The Love Song of J. Alfred Prufrock"

UNDERWATER SOUNDS RANGE FROM FUNNY to gorgeous to, honestly, kind of boring to humans. Yet to my ear, the most beautiful sound is the trill of the bearded seal. These seals' pure whistles sweep up and down, crisscrossing each other on the spectrogram. The seals are semi-aquatic and spend a lot of time on rocky haul-outs, where they certainly make a lot of sounds. But seals are also acoustic beneath the waves.

Bearded seals are an exclusively Arctic species, and in the summer of 2022, William Halliday was headed up to "seal camp" to study seals, including bearded seals, in situ.

Halliday is the Arctic acoustics program lead of the nonprofit Wildlife Conservation Society of Canada (and the author of the helpful paper that guided me to the plainfin midshipman). Halliday has the vigorous beard, good-natured countenance, and practical, unflappable focus of someone who has planned and survived a lot of logistically complex fieldwork. He's one of a handful of acoustics researchers who travel north to measure the frigid seas directly. This can be . . . tricky.

The Canadian Arctic Archipelago is nearly 1.5 million square kilometers, with most residents living in its small communities. Three of Canada's territories comprise its Arctic. The Yukon, with just 100 kilometers of coastline abuts Alaska west of the archipelago. Above Alaska and the Yukon lies the Beaufort Sea. Heading east, the Northwest Territories comprise the westernmost islands in the archipelago and the central swath of the Canadian continental north. The easternmost territory of Nunavut stretches across nearly to Greenland.

Halliday thought seals were under-researched in the Arctic compared to the toothed belugas and narwhal, and the baleen bowheads, an extremely long-lived Arctic species with an elaborate song. Several seals, including ringed seals, bearded seals, and harp seals, are major predators of fish and invertebrates and are themselves important food for killer whales. Polar food webs tend to be simpler than those farther south, with fewer species, and seals' important place in the food web is just one of many reasons they, and their reactions to sound, are worth studying.

"Seal camp" is 30 kilometers outside of Ulukhaktok, on Victoria Island in Canada's Northwest Territories. There Halliday and his students camp in tents, and ration their limited power from solar power and batteries, with a generator for cloudy days. The monetary cost of flying anything to the Arctic is high, so weight is restricted and dehydrated dinners are the standard. Local wildlife technicians come out with Halliday's group and armed wildlife monitors watch for any curious grizzly or polar bears in case they approach too closely.

In the past year Halliday had designed ship-playback experiments. He would tag a wild seal, then play something, like the sound of a ship, and see how it reacted. Halliday had headed up north in May, when the 2-meter-thick sea ice had begun to break up along the coast and seals gathered around the cracks. Then, he'd sunk hydrophone moorings rigged with cameras. It was crystal-clear water but dark beneath the ice, so visibility topped out at 10 meters—not great, but enough that he could watch their behavior. Now in August he would scout the ice-free ocean 20 kilometers around the camp for seals, put out nets to snag them, tag them, and then record their movements on the tags.

The first step was getting up north. This hadn't even been an option for the past several years: Covid-19 brought fears of outbreaks in remote Arctic areas days from medical treatment. Restrictions were very tight. Transport Canada banned non-essential travel from southern Canada up to the territories in 2020 and 2021, stopping Halliday from picking up the recorders he'd set out in the summer of 2019. He had scrambled to coordinate with the communities.

"I had all of the gear here at my home office," he says. "I set everything up, had it running, and all the ropes tied, and everything, and then shipped it up in a big box." Locals then dropped the hydrophones in about 30 meters of water, about 5 kilometers from shore. In the summer of 2021, they retrieved them and shipped them back. "And they both have great-looking data on them."

Halliday finally headed to seal camp in 2022 to try his first playback experiments. This involves a commercial flight to Inuvik, and then a charter flight to Ulukhaktok, and finally a 30-kilometer boat ride to camp. The charter flight is only an hour and a half, but it can cost thousands of dollars, weight is strictly monitored for safety, and the schedules can shift at any moment due to bad weather like fog. That was exactly what happened: Fog delayed the charter flight for a day and high winds delayed them a further two days once they reached Ulukhaktok. It took Halliday three days to reach camp where, despite precautions, one camp member fell ill and had to self-isolate in a tent.

Halliday assembled his crew and gear, determined to tag some seals.

But the seals were having none of it. For twelve days, they did not catch a single animal. To make matters worse, bad weather ate up half their field days, and winds twice gusted to 90 kph. The main canvas-wall tent blew over at one point. The tent fabric luffed so loudly that sleep was impossible.

Before he headed back Halliday followed one untagged seal on a foggy day. Despite the bad conditions he dropped a speaker and played some ship sounds, then watched as the seal swam away from the underwater speaker. It could be nothing but an anecdote. But it was promising.

Sure, Halliday could forego all this, extrapolate from studies on seals farther south, but there's no substitute for local data. The Arctic soundscape is completely different, generally much quieter than seas farther south. Animals here are used to hearing over larger distances than their southern, noise-hardened brethren, which means they may be responding to things farther away, make quieter sounds, or react more strongly to a novel noise than most of our data from other latitudes might suggest.

There is good evidence some Arctic species, such as beluga and narwhal, are much more sensitive to noise than southern species. In one study, an Arctic beluga reacted to the noise of an icebreaker 50 kilometers away, a sound barely above ambient levels. "It makes a lot of sense," Halliday says. If the beluga encountered maybe two ships a year, even a distant propeller might alarm it. Halliday has also found that belugas near the Mackenzie River Delta, north of the Yukon, decrease their vocalizing near ships.

The "zones of influence" still hold true, but the same sound means different things to different animals in different places. These variables are important to measure in situ because such models inform regulations, like how close ships, seismic surveying activities, or other noise sources can be to animals.

The Arctic is one of the last places on Earth where human sounds are relatively rare. Before this changes, Halliday wants to gather baseline acoustic data that, in most of the world's oceans, is now impossible. One major solution to underwater noise is good policy. Polar regions are a rare opportunity to design and enact such policy preemptively. Halliday spends days en route, sleeps in gales, and couriers hydrophones across Canada so that his playback and tagging experiments on whales and seals will collectively inform that policy.

So will his other, longest-term work: monitoring and recording ambient sound, ship noise, and fish and mammals in the icy waters. Halliday has a long-term recorder off the town of Sachs Harbour on Banks Island, Ulukhaktok farther east, and 13 locations scattered throughout the region as well as some recorders in Nunavut and one in the Chukchi Sea. He uses recorders powered by nine D batteries, which sit in the water for a year at a time. For the recorders near Ulukhaktok and Sachs Harbour, he likes to drop them in about 30 meters of water, near enough to shore that local vessels can safely reach them but deep enough that shifting, cracking sheets of ice won't dislodge them.

The recorder is sunk with a weight and an acoustic release, a clip that discharges the floating recorder at a certain signal. (Weights for such hydrophones range from buckets of cement to old train wheels.) Off Ulukhaktok, near seal camp, Halliday has year-round ambient sound dating back to 2016, and he can describe an acoustic year.

At New Year, the sea is dark and ice-covered, quiet but not silent. Ice creaks. Fish grunt quietly as bearded seals trill. When the ice breaks up around May, it booms and cracks, and the hiss of wind and waves rises. Belugas begin to call, and bowhead whales, too, as bearded seals fall quieter. Whale calls peak in July and August, as vessel noise peaks too. Motorboats fill about 5 percent of July's recordings. In August, 10 percent contain the sound of a larger vessel within 10 kilometers.

A few cruise ships ply the Arctic alongside Coast Guard boats, military ships, and farther east, near Baffin Island, commercial fishing boats. Ships supply Arctic communities with food, vehicles, clothing. Ships make the loudest summer noises, more than 120 decibels at their peak. The bowheads and belugas call less when it's noisy, or maybe their voices just carry better when it's quiet.

Ice covers the sea again by November, and the whales go quiet as they migrate to the Bering Sea. Bearded seals begin to call again. Occasionally, a snowmobile buzzes over the ice above. Sound levels are driven by ice, wind, and waves, and in the summer by vessels.

Halliday hears fish too. They grunt or croak on his recordings every month except December. They peak in August and again in March. The August fish make shorter, lower-frequency sounds, which could mean multiple species, or two different calls of the same species. He doesn't know for certain, but some of these sounds closely resemble those of the cod *Boreagadus saida*. (Some jurisdictions call this fish the polar cod, others the Arctic cod: We'll call it polar cod here.)

Polar cod eat zooplankton like fatty copepods. These cod themselves are also the preferred dinner of beluga, narwhal, and seals. They are widespread and a key link in the food web. So understanding their ecology matters.

For a long time, polar cod did not appear to make sounds. No one knew if they couldn't, or just wouldn't. Listening to them in situ was challenging, but they are very hard to keep in captivity because they need extremely cold water. If water temperature rises more than a few degrees above freezing, the fish die.

In 2018 Francis Juanes heard that Matthew Gilbert, a fisheries student at the University of British Columbia (UBC), a short hop across the Strait of Georgia from Victoria, was keeping some polar cod in a tank. Juanes Lab member and research associate Amalis Riera had worked with Rodney Rountree, and read Marie Fish and others. She was interested in fish sounds but knew that a real "audition" meant you had to isolate it and record the animal—"sound-truthing" as Rountree called it. So Riera headed to UBC to audition the polar cod. The tank sat in a small, isolated room stacked with cooling equipment whose mechanical noise would have made any cod sounds almost impossible to hear. She turned the equipment off just briefly enough to record a faint grunt.

Encouraged, she headed to the Hatfield Marine Sciences Center in Oregon, which had also managed to rear a few polar cod. There, she got more preliminary grunts, the first records of the acoustics of a keystone fish species in a critical region.

Why go to this trouble when we have lots of recordings of the cod's Atlantic cousins? When any fish make sounds, they make themselves obvious and vulnerable to predators such as hungry seals—but if they are too quiet, they risk not achieving what the sound was for in the first place. The Atlantic and Arctic Oceans have very different soundscapes. These different sounds carry differently. What the fish is achieving and risking when it grunts is different too—not to mention how much a noise like that of a ship might mask its call. These are big life-history questions for a fish that's such a central species, and yet more evidence that science cannot just be ported from one region to another. Data must be gathered on the ground. And the need for accurate data is high. For change is coming.

THE ARCTIC IS expected to shift profoundly in the coming decades, driven by climate change and the retreat of thick multiyear-old sea ice. The region is warming four times faster than the global average, and sea ice cover reaches new lows nearly every year. The Intergovernmental Panel for Climate Change (IPCC) predicts an ice-free Arctic summer as early as 2050. Near Ulukhaktok, storms already come more frequently in August, energized by newly open water in the Beaufort Sea and Amundsen Gulf. Halliday learned this firsthand in 2022, and is shifting the dates of his fieldwork accordingly.

Insofar as threats can be considered separately, the threat of noise pales beside the cascading effects of climate change, pollution, and overfishing. Yet as work in Tadoussac, Cook Inlet, and elsewhere shows, sound can compound existing problems and add stress to already vulnerable species. So, understanding this pending change is invaluable.

More open water will admit more ambient wind and waves, raising background sound levels. Species might have to call louder to be heard. That might be risky for prey hoping to escape the notice of hungry seals.

More sunlight hitting ice-free water feeds more phytoplankton and builds nutrients. With warming water and circulation changes, new species will arrive, bringing new voices. Killer whales move through the Bering Strait into the Chukchi Sea in the summer, and as ice thins, they already are arriving sooner, and swimming farther north, than in colder times. Killer whales also move farther into Hudson Bay at different times of year, following prey such as seals and bowheads.

Atlantic cod are moving northward even as polar cod, particularly sensitive juveniles, find their necessary cold water squeezed ever northward. Will their predators follow them, or develop a taste for the Atlantic variety? Off the west coast of Greenland, warming water shifts the relative proportion of different copepod species. All levels of Arctic food webs are either poised for change or already changing. Soon, the dialects of killer whales, and grunts of Atlantic cod cousins, may ring over the hiss of open waves where once only beluga whales whispered, or polar cod murmured, beneath ice.

Humans have lived in the Arctic, in what is now called Canada, for millennia. It is the traditional home of several Indigenous peoples, but most of the land today is that of the nine Inuit groups. Inuit call their lands Inuit Nunangat. These stretch from present-day Russia across Canada to Greenland, in four regions called Inuvialuit (northern Yukon and Northwest Territories); Nunavut (where Ulukhaktok sits); Nunavik (Northern Quebec); and Nunatsiavut (Northern Labrador). There are about 70,000 Inuit in Canada, most of whom live in several dozen communities throughout these lands. Fewer than 1 percent of Canada's population lives there, but it makes up more than half of Canada's coastline. These communities rely on fishing and hunting, including of bowhead whales, narwhal, beluga, and seals, and they rely on sea ice in many places for transportation.

But ice-free water and untapped natural resources means a likely rise in shipping, tourism, and industry to the area. There's the distant promise of shipping lanes: The fabled Northwest Passage promises a fast shipping route from Europe to Asia but is only open a few months per year, so ships either head below Cape Horn or transit the Suez Canal or the Panama Canal. But an ice-free Arctic shipping route might draw more commercial ships, the kind that caused those noisiest spikes on Halliday's existing recordings. The passage wends through the archipelago not far north of the mainland, past Halliday's recorders at Ulukhaktok and Sachs Harbour, making his baseline data there immensely valuable.

But realistically, the passage will stutter into use slowly. The biggest source of ship noise now and in the immediate future is vessel traffic to mines on Baffin Island, which holds rich mineral deposits. Bunkers carry iron ore from this large island north of Quebec, notably from a mine at Mary River, while tankers bring in fuel. The waters can see multiple ships a day.

Halliday's work informs more than policy. It also helps locals deal with things like mining permits, environmental assessments, even legal battles.

Canada has the longest marine coastline in the world. It borders three oceans: Pacific, Atlantic, and Arctic. Canada, at least politically, presides over a huge swath of the Arctic Sea and its coast. "So doing something about [the coming changes] now before it gets bad is kind of the key," Halliday says. If we can set up protections now, he says, like properly designed shipping corridors and marine protected areas, we stand a better chance of preserving baselines—far better than trying to develop such interventions after the fact.

Shipping is one of the few industries (aviation is another) with an overarching international body, in this case the International Maritime Organization (IMO). The IMO, as part of the United Nations, is a panel of member states. When a majority agree something should happen, each state then makes its own regulations. The IMO in 2014 put forward non-binding guidelines on underwater noise. But to date, the only jurisdiction with enforceable regulations on underwater noise is the EU, via its Marine Strategy Framework Directive. In 2016, Canada launched the CAD $1.5 billion Ocean Protections Plan (OPP), which includes the non-binding Ocean Noise Strategy, expected to go into effect in 2023, that lays out how to regulate underwater noise. In the United States, NOAA also has an Ocean Noise Strategy.

Noise policy needs standardized units. In 2009 the American National Standards Institute (ANSI) put out guidelines for measuring the sound of ships. The International Organization for Standardization (ISO) has standards on acoustic terminology, measuring ship noise in deep water, and they're developing more, including ship noise in shallow water and the measurement of sonar.

Aside from shipping, another noise source in the Arctic is seismic air guns, which have been used extensively off the north shore of Alaska and in the Beaufort Sea. Seismic air-gun sound can travel 100 kilometers in the Arctic. "Some studies have even documented them up to one thousand kilometers away," Halliday says. "Very far, very loud signals. And because they're such low-frequency signals, if they get caught up in the right kind of acoustic-propagation ducts, then they can go very far." There has, however, been a moratorium on exploration with no seismic surveys in the Beaufort since 2017. Exploration off the west coast of Greenland by the Danish yielded little, and was ramped down.

In 2016, the community of Clyde River on Baffin Island vigorously protested a proposal for seismic surveying in the nearby waters to image the rocks beneath the seabed, in the search for oil and gas. The proposal was approved by Canada's National Energy Board, but the community and a range of national and international supporters appealed the decision. They cited a lack of consultation, and lack of knowledge about the danger of these loud noises to marine animals, including species of seals, whales, and walrus, that are hunted by and have great cultural significance to, the local people. The case made its way to Canada's Supreme Court, where the approval for testing was eventually withdrawn, thanks in part to concerns about underwater noise.

Good policy requires good data, including baseline measurements. But do we actually have such data farther south?

IN 1963, AT William Tavolga's conference, Gordon Wenz presented a paper entitled "Acoustic Ambient Noise in the Ocean: Spectra and Sources." Scientific papers are not known for being good escapist reading, but I've come back to this one again and again because it encapsulates all the different noise sources he heard in the ocean. (And, well, I just think it's beautiful.)

Wenz describes these sounds in the ocean with attention and care: "Wind-dependent bubble and spray noise," categorizing the noise that comes from various sea states—from "mirror-like" to "heaped-up sea, blown spray, streaks" and even spindrift, where the wind blows the crests of breaking waves, which always makes me think of Eliot's verse "Combing the white hair of the waves blown back / When the wind blows the water white and black."

I love to revisit this paper now and then, though there are many more recent or updated figures. Yet even in the 1960s this figure shows industrial man-made noise. Lines arc across the mid-to-low frequencies, for "heavy traffic noise"; "usual traffic noise: shallow water"; "usual traffic noise: deep water." Wenz says traffic noise in the deep sea can easily reverberate over 1,000 kilometers. Outside of the Arctic, do we have any sense of what a pristine sea might sound like?

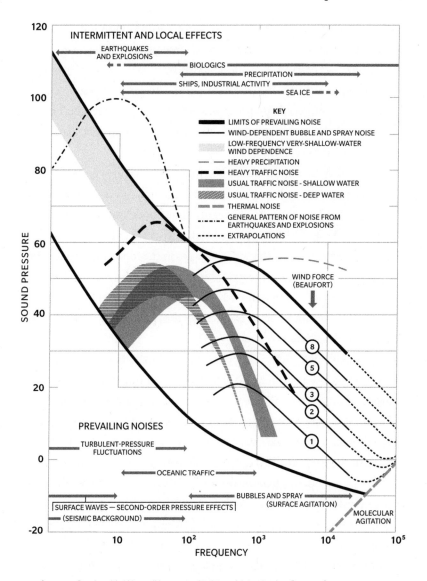

Source: Gordon M. Wenz, "Acoustic Ambient Noise in the Ocean: Spectra and Sources," *Journal of the Acoustical Society of America* 34, no. 12 (December 1962), 1936-1956 DOI: https://doi.org/10.1121/1.1909155.

In most places, we don't know what the sea would sound like when we are not in it save a few exceptional circumstances, which unfortunately, often accompany exceptional tragedy.

On September 11, 2001, following the terrorist attacks on New York City, an unprecedented halt was called to global travel and transportation. Planes were grounded: Skies were contrail-less. And ship traffic dropped too.

In the Bay of Fundy, where North Atlantic right whales come to feed and nurse their young in the late summer, lies a shipping lane. Scientists at the New England Aquarium had been studying the whales there since 1980. The September 11 attacks grounded everything, but the team managed to collect some samples in the following days, during which the sea was much quieter. Ship transits in late August averaged four or five a day: on September 12 and 13, they dropped to one or two. Noise in the ship-dominated frequencies dropped by 6 decibels. And so did the whales' stress hormones.

Certainly, such studies can't confirm if animals are stressed from the noise, or just the presence, of ships. Maybe ships stress whales just by existing, their monstrous bellies gliding across the sky. Still, until March of 2020, the study was one of the only direct examples of how marine animals behave when we shut up.

Then the coronavirus lockdowns happened.

Covid disrupted research worldwide. It prevented Steve Simpson from traveling to Lizard Island. Socially distanced boat trips forced smaller crews and canceled fieldwork. Many researchers lost a year, often two, of data.

Researcher Christian Rutz coined the term "anthropause" in a 2020 paper that suggested the lockdowns were an opportunity to learn unprecedented amounts about how wildlife behaved without human interference. And it did seem, at least to many urbanites, that wildlife was having a field day (though skyrocketing sightings of animals in cities may have had more to do with our lockdown observation skills than any real change in the animals' behavior). In the ocean, it seemed like a chance to gather priceless baseline data, including acoustic data, of a human-free environment, an opportunity that would never have been possible otherwise.

The promise of the anthropause, was, in some places, fulfilled. Researcher Matt Pine happened to be in New Zealand and had some recorders in the ocean off Auckland in the Hauraki Gulf. When the New Zealand government mandated lockdown they forbade people to go boating and the harbor went quiet. Pine then heard everything from invertebrates crawling on the recorder to dolphins vocalizing more often than before; their communication space expanded from under 3 kilometers to over 4.

In Glacier Bay National Park, a humpback whale habitat and ubiquitous cruise-ship destination, the sudden lack of ships made it quiet enough that National Parks Service hydrophones monitored on shore picked up the soft calls of mother and baby humpbacks, murmuring to keep close.

But the story was different elsewhere. In Sarasota Bay, Florida, people were allowed to quarantine on boats. Noise at two sites either stayed the same or rose by 80 percent, respectively. Off the coast of Vancouver Island, locking down on sailboats and other motored pleasure craft was allowed. Boat sales spiked.

Preliminary anthropause numbers from the Ocean Networks Canada hydrophone networks, NEPTUNE and VENUS showed that depending on shipping-lane location relative to the nodes, for the first few months of lockdowns, sound in the deep ocean remained the same or dropped by about 1.5 decibels, and in the Strait of Georgia, depending on which node you listened at, boat sound dropped by between about 2.5 decibels to almost 7 decibels. This corresponded with a drop of about 13 percent in container capacity through the Port of Vancouver, and about 21 percent reduction in ship transits.

However, ship traffic rebounded—and then some. Global container trade hit unprecedented levels by 2021, spurred in part by a flurry of online purchases. This compounded with labor issues due to the pandemic itself, led to weeklong traffic jams at the ports of L.A. and Long Beach. Many hoped the anthropause would yield data to prove that animals heard or communicated better in our absence.

In some locations, where opportunity presented itself, it worked out this way. But more often, the quieting was only a few decibels. And as more research emerges, even the most dramatic examples largely illustrate what we kind of already knew. So, the anthropause didn't provide us with the slew of data we hoped. But perhaps there's some other strategies to survey, explore, and understand marine bioacoustics more quickly than we have to date, as climate change and industrial development accelerate ocean change. I came across one such strategy that is gaining traction, while learning about Kieran Cox's kelp project: the measurement of the soundscape.

IN 1968, BERNIE KRAUSE was tapped by Warner Music to record an album called *In a Wild Sanctuary*. That meant going outside and recording sounds in nature. With no idea of what this would entail, or how best to do it, he headed out to the Muir Woods, north of San Francisco, figuring he would record the sounds of the forest as they were.

"Because I came to it from a musical perspective, I was not [supposed] to separate individual sounds out of context, but to capture the whole," he says. It would be a more holistic context, the natural "orchestra" of wildlife sounds in that habitat. He could hear the world open up, the rush of raven wings, the whisper of leaves rustling in the tree canopy. The trickling of water in a nearby stream. He'd never heard anything like it before.

He was captivated by how peaceful these soundscapes made him feel. "I just decided that's what I wanted to do the rest of my life," he says. Krause was, at the time, also a synthesizer programmer, working as a session player in windowless studios for acts like the Doors and the Beach Boys, and growing disillusioned with the scene. After *In a Wild Sanctuary* came out, Krause kept working in music, including for the film *Apocalypse Now*, but began transitioning to working on natural sounds.

The scientific approach to sounds above or below the waves is, typically, zeroing in on a single species, identifying one voice or another and not considering the bioacoustic whole. "To me, the approach has always seemed a bit like trying to understand the magnificence of Beethoven's Fifth Symphony by abstracting the sound of a single violin player out of the context of the orchestra and hearing only that one part," he writes in his book *Voices of the Wild*. Krause's musical background drew him to consider the entire suite of sounds. The acoustic fingerprint of a place.

If you've heard about "soundscapes" in the past decade or so, you likely have Bernie Krause to thank. Krause didn't invent the term: It was coined by Buckminster Fuller in a 1966 essay and used in relationship to urban planning in a paper in 1969 by Michael Southworth. But the late Canadian composer Murray Schaffer, with whom Krause worked since the 1980s, is credited with popularizing the term as it might relate to the music of natural places. Krause remembers around 1977, "he was using it sort of in terms of humans using it. For musical expression."

In the late 1970s, Krause went back to school for a PhD. He began at MIT and finished at the Union Institute, in Cincinnati, in 1981. He briefly studied marine bioacoustics, even traveling to Johnstone Strait in British Columbia and comparing the vocalizations of killer whales in the wild to those from the same wild pod but now captive in California (the captive whales were at a theme park called Marine World Africa USA). But Krause eventually turned to terrestrial soundscapes and began to develop his ideas. One was the acoustic niche hypothesis.

In 1983, on a recording trip to Kenya, Krause was lying awake late at night. He was exhausted, having been up for some thirty hours. In his slightly loopy state, he was listening over the day's recordings. As he did, he started to hear the suite of sounds not as a random by-product of the ecosystem but as structured, as organized in some way.

"When I got back to California and got back into the studio, there were these primitive spectrograms that were available at that time," he says. "And I ran the material through the spectrogram-panel application. And sure enough, I mean, the sounds of the insects have reformed in one niche, the night birds and frogs, mammals and so on. Each occupying their own frequency and temporal bandwidth." He suspected the animals had evolved their sounds to occupy areas of the sound spectrum not filled by other organisms, so that each species' sounds didn't interfere with one another.

He says that others around him, used to focusing on the sounds of one species at a time, weren't receptive at first to the idea that animal sounds had this relationship to each other. "If you want to understand how the birds are singing in relationship to the other sound around them, you've got to see the whole fabric of sound," he says. The term "niche" had been used in ecology since the 1950s, when it was used to describe how each animal found its place in the ecosystem. Krause and his colleagues now began to use the term "acoustic niche," or how each animal made sounds in its own unique time, place, and frequency, so it fit into a whole.

Working with Stuart Gage, Krause coined the term for what he was listening to: *biophony*—literally, the sounds of life, the collective sound produced by all organisms at one place and time. They joined these together with *anthrophony* (human-generated sounds, further divided into controlled sound like music, language, or theatre, and chaotic or incoherent signals called noise) and *geophony* (non-biological natural sounds like wind in the trees, waves at the ocean shore, and the movement of Earth). In Krause's framework, natural soundscapes contain mixtures of biophony and geophony.

Yet we've detected all three elements in the ocean since Wenz's first 1960s graphs.

Krause and his colleagues' framework of "the soundscape" has become a popular way to study underwater sound. Many acousticians now study the sounds, not of a single species, out of context, but of a place's entire acoustic signature. One appeal is that we may be able to learn a lot of information with a relatively simple measurement: listening. Fieldwork and monitoring species in situ is time consuming, expensive, logistically challenging, and occasionally dangerous. By studying a soundscape, we may be able to do a lot of this just by listening. We learn more, and with less effort and expense.

This is particularly appealing in remote or challenging ecosystems. One place where this could be of great use right now is the very deep sea, which, at this moment, scientists are working to characterize its baseline ecology before human activity and industry intrude.

For all we study sounds in the sea, the vast majority of species and research sites are near the surface, or near the coast. But of the vast, deep seabed (broadly, anywhere more than 200 meters deep, where light is diminished or gone), only about 20 percent has been mapped, let alone studied.

In some very deep places, more than 3,000 meters or so, the seabed is scattered with potato-sized lumps that, when brought to the surface, look unprepossessing, like rust (or, frankly, I think dog poop). But these nodules are rich in minerals precipitated out of the seawater millimeter by millimeter over millions of years in this very deep water. They contain manganese, an ingredient in steel, but also cobalt, nickel, copper, and rare-earth minerals that are used in electronics such as cellphones.

The idea that these polymetallic nodules might be profitable to mine was first raised in the 1960s. By the early 2000s, the price of manganese and other rare-earth minerals was rising and mining interests began to assemble, seeking to exploit this untapped resource.

The seabed is governed by the International Seabed Authority, a branch of the United Nations, which as of 2023 is still struggling to put a mining code in place to regulate private deep-sea mining companies that want to get started harvesting these nodules. Because they form in deep water, many of these metallic nodules exist in fields on the seafloor in the depths of the Pacific Ocean, around Pacific Island states such as Kiribati. In a region called the Clarion-Clipperton Zone, southeast of Hawaii, there's an estimated 21.1 billion dry tons of manganese nodules. Many of the island governments in these regions are deeply divided on how this potential new industry may help or harm them. And we know almost nothing about deep-sea ecology, what harm from mining activities might look like, or how we should best measure the impact of future mining operations.

In such a challenging environment, measuring the soundscape could give us a lot of data quickly. We wouldn't know what we were listening to, but we could get a baseline reading, and monitor the acoustics of the deep sea for any changes. Researchers have measured the soundscape of several deep-sea locations off the coast of Japan, in the hopes of beginning to characterize the soundscape of the deep sea at different types of locations: a seamount, a 5-kilometer-deep abyss, and two other shallower spots. The shallow spots hummed with ship noise. The deep site was quieter, even, than sound recorded from the bottom of the Mariana Trench, the deepest point on Earth (where ship noise is still quite audible). One site had a hydrothermal vent, a crack in the ocean floor where hot water from within the crust plumes up into the sea, and this site growled with low-frequency sound caused by water thrumming through the rocks. Each site had a different soundscape.

It's now well-known that the sound of reefs helps guide animals in the sea's upper reaches. The same phenomenon may very well occur in the deep sea. Rumbling vents attract clusters of deep-sea creatures, forming oases of life in the relative desert of the seabed. We know that some of these animals also have a larval phase. We don't know if deep-sea vents and ridges hum a homing song to these larvae the way that coastal reefs do. But if so, it suggests these sounds reflect habitat quality, and thus we could listen to the noise of the communities around deep vents and chimneys as a litmus test of ecosystem health.

Other unique, or entirely new, ecosystems can be understood at least in part through their sounds too. Glass sponges were thought to have been long extinct, until they were discovered in the 1980s in relatively deep (for a reef) waters around Vancouver Island. They are what they sound like: delicate, white, porcelain-like sponges. They live in relatively deep water and provide habitat and a gathering place for animals at those depths.

William Halliday was among those who recorded the glass reef's soundscape. In 2016, he and colleagues listened to the sounds among the glassine structures in about 90 meters of water, just north of Active Pass. They counted 41 fish sounds in the reef (19 grunts and 22 knocks); at the edge of the reef, 120 sounds (36 grunts, 84 knocks); and off the reef only 7 knocks and no grunts. It was, in fact, its own unique soundscape, a unique assemblage of animals. These sounds told the research team that the reefs are likely a habitat for fish, a gathering place, and a unique community. They also heard peaks in noise from ship propellers cavitating and vessels passing at all hours of the day.

The soundscape framework is good at some things, and not at others. It's good at describing sound levels. It can give a sense of how groups of animals partition themselves both acoustically and in time. It can tell us that there's a reef chorus in the evenings, or if two species call at different times. It can be a monitoring tool and a benchmark for change. If soundscape features change, it could be an early warning of problems: just as louder reefs draw more larvae than quieter ones. And soundscapes can give us at least a partial census of species.

To that end, many scientists are now building on this framework and developing soundscape indices. Soundscapes can be measured on how periodic they are—for example, do patterns repeat daily or seasonally? Are sounds impulsive? Are most sounds in one place the same, or do they vary? Such soundscape metrics can be useful—but there are limits. As Marie Fish and others have found, without a catalog of what species makes what sound, we don't know who's there. For example, the species on the glass reef in British Columbia were only identified because of a visual survey. Soundscape measurements can't tally the number of animals living in a place. Standing on the dock and listening to the midshipmen, I knew there were fish along the shore, but I couldn't tell if it was fifty loud males or a hundred soft-spoken ones.

Rodney Rountree, the fish listener and adjunct professor at the University of Victoria who introduced me to cusk-eels in Cape Cod, is more equivocal about the value of soundscape indices and metrics in ecosystems we don't understand fully. "I've been a skeptic of that for some time," he says. "I would much prefer to see [. . .] actual sounds identified." He says that without old-school field observations of the species and relationships in an area, the numbers can be meaningless or even misleading. That's sometimes better than nothing when funding is short and regulators are clamoring for guidance. But he thinks that's unfortunate. For a long time he has believed there must be a catalog of marine animals' sounds, so indices will be grounded in knowledge. And just as some scientists are hurriedly developing the soundscape framework in advance of climate change and industrial development, others are racing to catalogue and database individual marine species' sounds—especially fish and invertebrates, whose sounds are less well-known and documented than those of mammals.

When Rountree first got his hands on a copy of Marie Poland Fish and Mowbray's book, he noticed that the original printing of the book had included a cassette tape of the sounds it describes. Fish's alma mater, the University of Rhode Island, didn't have the tape. The analog recordings, Rountree learned, were magnetic reel or cassette tapes, but after a fire in 1959 destroyed Fish's laboratory, things had been moved around. Marie Fish herself had retired in 1966 and had passed away in 1989. Rountree envisioned reviving Marie Fish's efforts to have a library or a database, some master field guide to fish sounds. It would help researchers collaborate, and it would inform people who, like sonar operators, were listening underwater for one reason or another. It would help fisheries track their species, help biologists build an understanding of acoustic indices and the behavioral ecology of fish. But Marie Fish's archive wasn't easily available, and it still only covered a fraction of species.

Rountree finally tracked down a man named Paul Perkins, who had been a sonar technician in the navy and made some of the first recordings of gray whales as well as contributed to the fish archive. Perkins had used Fish's work to distinguish biological sounds from enemy subs and had helped train sonar techs in the Bahamas. Rountree also spoke with others who had worked with Fish, and eventually discovered that Fish's original tapes were on pallets in an old warehouse.

Rountree was aghast. Here was the first comprehensive collection of fish recordings ever made (plus some seal and whale sounds), moldering away in obscurity. He believed such records should be archived, not only for their historical value but because it was at least the beginning of a much-needed archive. So, he got the tapes. He digitized the original cassette, and then worked with Cornell University to rescue the historical recordings and digitize those.

"We need to build a new library," he says. "We need to have systematic catalogs of fish sounds and other animal sounds."

WHEN AUDREY LOOBY graduated from the University of Southern California with an undergraduate degree in environmental studies, specializing in community ecology, she wanted to pivot to studying soundscapes but didn't have the experience—the scientific street cred—to get funding in the field. So, she headed to graduate school at the University of Florida and continued studying aquatic ecology. Then, one day, a mutual friend introduced her to Kieran Cox. Over beers, Cox told her he was studying soundscape ecology for his PhD up in Victoria and offered to help her with any questions she might have. Looby had a basic one: How many fish made sounds?

If only a handful of fish sing, that's different than if they all do, or if only some families do, or only in certain habitats. Such differences would change how sound was worked into ecosystem models. It would affect regulations for protected areas as well as our understanding of how and why fish evolved to make sounds in the first place.

The most common figure Looby found for the number of soniferous fish species was 800. But when she traced this number, she discovered it came from an unpublished data set, which originally cited 520 or so, and had morphed through time, like a statistical game of telephone. It was scientifically questionable. Cox confirmed that since Marie Fish, through Rodney Rountree, to the present, there was no tally, or database, of fish sound.

So Looby decided she would build it. In her spare time.

Working with Cox and Rountree, she began searching all of the published scientific literature for confirmed fish sounds and their descriptions. "Everyone thought I was weird," Looby says, "but it was honestly super enjoyable." It gave her a crash course in fish acoustic biology and scientific street cred.

Looby examined 2,943 references from 60 countries in 11 languages, including Japanese, Indonesian, and Russian, going back to 1874. (It helped that she could read French and Spanish.) She concluded that 989 species of fish, so far, are known to make sounds. But that's just fish that have auditioned. The number grows each year.

In 2021, Looby, Cox, Rountree, and their collaborators launched FishSounds.net, an open-source database searchable by recording, region, species, and even the collector of the sounds. It contains Marie Fish's recordings. Rodney Rountree has donated his extensive collection. Tony Hawkins sent recordings from his 1960s studies in Scotland.

From this data set, patterns have started to emerge. For example, soniferous fish are found all over the world, except (so far) in Antarctica. This underscores the broad importance of sound to fish in many different contexts, places, and species: not just to specialists.

Looby's personal favorite fish sounds are the "really nice, multitonal calls" from the toadfish family, which includes the plainfin midshipman. She likes it so much that the species is FishSounds's logo: an intense-looking toadfish with sound waves emanating from his puffed-up body.

Creating detailed databases of animals and their sounds is in some ways the opposite approach to studying a soundscape. Databases require listening to, say, individual fish: A soundscape approach looks at the whole. However, both approaches can be complementary as we race to learn more about the world of sound underwater.

I ask Cox about soundscapes, on the boat near the seal haul-out, and he describes one big advantage of having a simple metric, as some soundscape indices provide: communicating science to non-specialists. His playback experiment touches on many disciplines; kelp forest ecology, habitat loss, introduced pollutants. To translate his research between disciplines, or into policy, he needs a common language, as he witnessed when speaking with Members of Parliament and others on the Ocean Noise Strategy. Tools like indices, for example, help him explain his work to more people.

"I'm a bit critical of some of the indices work that gets done," he admits. But he's a realist. He likens indices to a currency. "Don't just try to convince people to love the things you love," he says, a sentiment that, as a science writer, both breaks my heart and resonates deeply. "I would never go to another country and expect them to take my currency. I would never sit down at a table with a politician and expect them to take my currency, which is love of nature." Yet it is also, as he is about to tell me, the soundscape of the kelp forest that inspired this very project. The morning shades into afternoon, as we float.

"You hear a lot when you dive," Cox says, watching his teammates' bubbles.

In fact, divers use sound often to coordinate, as their time underwater reveals that even human hearing can sometimes be of use. To get each other's attention in the murky depths, they'll bang on their air tanks. Cox will clap one hand onto a closed fist, making a sound his diving buddy can hear.

"Diving is such an odd thing for me," Cox says. "It's certainly the time I feel the most calm, and quiet, and relaxed."

I think I know what he means. My own restless thoughts quiet beneath the surface when I snorkel or dive. It always puts me in mind of a passage in Alex Garland's *The Beach*, in which the narrator tries to calm his mind by jumping in the ocean: "Underwater had always had the qualities of a refuge for me. Calming, blinding, deafening; a perfect escape. It worked, too, enveloping me in anonymous coolness, but in an unavoidably temporary way. Without gills I had to keep surfacing, and as soon as I surfaced my mind resumed its circular debates."

It was while diving with a friend that Cox got the idea for the kelp-sound attenuation project. He was sitting in a very kelpy dive site called Ogden Point in downtown Victoria on the southern tip of Vancouver Island. The Ogden Point breakwater is an 800-meter dogleg of stone blocks that juts into the Strait of Juan de Fuca, minutes from the city's downtown. The breakwater's picturesque length is often choked with tourists, joggers, and dog walkers, but it's fringed by a spectacular bull-kelp forest, which makes it a great cold-water dive site. Seals twist through 20-meter stalks. Giant Pacific octopuses, ghostly white plumose anemones, wolf eels, rockfish, nudibranchs, and dozens more animals cluster in the kelp.

"I jumped in," Cox recalls, "and I landed right in this kelp, and the light was coming through the kelp, with all these tube snouts and black rockfish . . . And I just noticed that kind of *stillness* of it all." He likens it to the stillness of big terrestrial forests, not so much an obvious hush as a subtle change in sound quality. It would take an experienced diver, who could easily hover still and enjoy the habitat, to notice. The longer he sat, the more Cox, in a rare moment of stillness, noticed the effect.

He was in the final year of his PhD and wondering about his postdoc. Soundscapes had never been his focus, but he'd always found them interesting. But drifting in the kelp, *I think this is a thing,* he remembers thinking in the forest hush, *and I think I can quantify it.* The stillness, he thought, was a conservation of the soundscape. As kelp habitat was lost, maybe this stillness was, too. He'd published a few papers on soundscapes by then. He'd also worked on habitat complexity and noticed sound was often a missing component. "I knew it was outside what was being asked," he says. "I knew what a novel question it was." The answers to such questions will lead, piece by piece, to meaningful policy.

The kelp study is thus a bit of leap, connecting disparate topics: soundscapes and surveys, ecology and acoustics. It's complex. But then, so is life.

COX'S KELP-FOREST HUSH, like Krause's first experience of biophony in the Muir Woods are all moments of acoustic transcendence, a new world revealed. Often these witnesses describe not a single voice but the all-encompassing-ness of a, well, a soundscape. Just like simple metrics and indices, awe is a powerful motivator, and many are working to bring the acoustic world beneath the waves to the public using technology, science—and art.

Krause is over eighty now. He could hear up to 18 kHz until he was seventy, but over the last decade, his hearing curve narrowed, down to peak at 7 kHz. Then one morning, the hearing was gone in his right ear. Just gone.

"I can't hear space, and I can't hear depth," he says. "That's what's really hard for me. But I learned to read spectrograms. And so because I can read the spectrogram, I have a really good sense of what the space is like. Even though I'm not really hearing it."

For the last ten or twelve years, he says, "most of my time has been devoted to transforming the data that I have in my archive into works of art, and sound art." When I catch him in 2022, his sound installation, *The Great Animal Orchestra*, in Paris, has been seen by more than a million people since its premiere in 2016. It's been shown in London, Milan, Shanghai, Seoul, and many other cities around the world.

"Art" is another of those words, like "language" or "culture," for which a description is a bit of a fool's errand, particularly with respect to the natural world. But *Songs of the Humpback Whale* was a bestselling album, and since then, the sounds of the sea have inspired not only researchers but artists. To name but a few recent pieces: the duo Luftwerk created a haunting sound installation in Chicago, using sound from a disintegrating iceberg, inspired by the cracking-off of a 5,800-square-kilometer section of Antarctica's Larsen C Ice Shelf in 2017. Artist Jana Winderen created a piece to accompany a 2021 paper in the journal *Science* on the soundscape of the "anthropocene ocean."

This book has been a scientific journey. But at the edges of the science, I've been delighted to find art, works of creativity and outreach. If the purpose of art is to connect, then bringing the ocean's sounds into public spaces and into private earbuds might be one of the most profound ways we can bring our human ears underwater.

IN THE EARLY morning before dawn I pack a bag and make coffee in a mason jar. I'm in a rush to catch the exact moment when the moon—in October, the Hunter's Moon—is full, which my smartphone has told me will happen today at 7:57 A.M. But the purpose of my trip is more than simple moon-viewing.

I drive the empty rain-slick streets of Victoria downhill toward the ocean, leapfrogging through the streetlights with intrepid early cyclists and lumbering buses, to the breakwater at Ogden Point. I was worried it would be dark, or I would be alone, but in the gloaming I see old women speed walking, young women jogging, men in tweed looking out to sea. I hear the skirl of herring gulls overhead.

I thumb my phone to radioamnion.net and wait to hear the sounds as I start out along the breakwater, the doglegged concrete path atop a wall of boulders jutting several hundred meters into the Pacific Ocean. At its farthest end squats a red-and-white lighthouse.

If I were to swim 300 kilometers offshore from where I sit, and then 2.7 kilometers straight down, I'd find the Cascadia Basin node of Ocean Networks Canada's NEPTUNE cabled seafloor observing array. Each node is a cluster of instruments: Scientists from all over the world can set up their gear there to take oceanographic measurements. Amid the fin whale–sensing geophones and the wind and rain hiss–sensing hydrophones, is a demonstration neutrino observatory run by German physicists.

Neutrinos are particles that stream in from outer space, and they are of great interest to astrophysicists, but they are difficult to detect. The easiest way is to go very deep down under the ground—like in mine shafts—or under the sea—where almost nothing but the neutrinos can penetrate. At the moment, part of the neutrino observatory's instrumentation is sending out very tiny pulses of light.

This is an art installation called *Radio Amnion,* spearheaded by ONC in cooperation with current artist-in-residence Jol Thoms and the German team. Each month, on the full moon, the neutrino observatory will "broadcast" a sound piece from a different artist into the sea, and simultaneously online for anyone to listen in.

These light waves blinked out down there have been transformed from sound: the sound of the performance I am listening to, right now. As I absorb the piece, it is also being blinked as light into the sea itself, just offshore. I've been tuning in each month. I can imagine that tiny pulses of light in the depths might be problematic for some of the instruments, and I'm not sure it's returning any "useful" data. But I think it's beautiful.

The past few full-moon broadcasts have been musical pieces, from abstract soundscapes to low, dulcet tones. For this month's full moon, the broadcast is a spoken-word piece by artist Caitlin Berrigan, a twenty-two-minute story entitled "A Voice Becomes a Mirror Plane Becomes a Holohedral Wand." A clear, monotone female voice narrates this strange and beautiful story. It quietly resonates from where I've nested my phone in my pocket.

It is the story of a creature, a woman, who dives to harvest shellfish for her community of "mothers"—and who must dive deeper and deeper to retrieve them as the oceans warm and the algae die.

The early morning world around me is blue and gray. Two container ships sit at anchor on Constance Bank, out in the mist. One has a string of white lights all round its railing, like Christmas. AIS data on quickly thumbed free tracking websites tell me exactly who I'm looking at. Both the MSC *Vega* and the *Maliakos* are Liberian-flagged, anchored here waiting for a berth to unload consumer goods. *Maliakos,* the nearer one with the lights, was in Busan, South Korea, ten days ago. She's more than 76 meters long and carries 4,300 TEU. MSC *Vega* also hails from Busan, two weeks back. She is huge, and her iron sides plunge 14 meters below the surface of the sea: nearly the height of a five-story building.

In the lee of the breakwater bobs a yellow pilot boat whose glowing strawberry-and-creme lights atop its tower tell me it's up and running.

I look for the moon without much hope, as the sky is fully clouded over. I use my Night Sky app, hold it up, and move it around the western horizon: there, farther north than I would have guessed, a tiny glowing ball just above the horizon (near Uranus, in Pisces, apparently).

The sea is calm and very clear this morning—Wenz would classify it as "rippled," verging on "mirrorlike." I can't imagine much wind or wave sound. When I look over the breakwater's railing, the variegated rocks waver in aquamarine. The tide is moving. The currents will soon be rushing through the cracks.

The story echoes my surroundings in a strange way. As the narrator says "ocean waves" I hear the bass rush of the surf's glassy swell breaking on the small beach beside the breakwater. As the voice intones, "The kelp beds are dead . . ." I look along the breakwater. Perhaps three meters out, bull kelp tangles and swirls on the surface, floats poking up like raised fists among hose-like coils of stalks.

This is the kelp forest whose shadowy hush inspired Cox to do his experiments in Bamfield. I've dived in this kelp forest myself, lugged scuba tanks out along the breakwater and fallen backward off the blocks. It's not the lush macro forest of Barkley Sound but groves of sleeker, slimmer bull kelp. As you sink down, 10, 12 meters, the kelp's round, gelled floats and streaming blades at the surface taper to slim stems, like hanging cables. It's tricky not to get tangled. White plumose anemones sprout from the encrusted stone blocks like the ghosts of 2-foot-high broccoli. A wolf eel glares above its underbite. A colorful nudibranch drifts by. A crack between two stone blocks that make up the breakwater is filled with something wrinkled and orange: a giant Pacific octopus the size of a beanbag chair, curled in its den, its uncanny eye glaring beneath a curled leg studded with perfectly round suckers. Twenty meters down, the kelp stalks end in holdfasts netted around rocks and boulders. Sea stars and urchins crust the twilit rocks as a harbor seal dives and twists with uncanny grace.

Still walking, I pass the spot where the BC Fish Sound projects tested its array, modeled after Rodney Rountree's greenhouse, as in my phone, the story continues. One day, our narrator dives very deep and discovers a hydrothermal vent and a forest of metallic nodules. She carries them up to the surface.

Right now, with the gulls and my own footsteps, I hear the broadband rush of noise from the docks opposite the breakwater, where people are already working. The beep of a truck backing up. I hear a small plane somewhere overhead, and the chop of a helicopter taking off from the helipad opposite the warehouses. It drowns out the story for a little bit, even with noise-canceling headphones.

I find a small stairway to the lower rocks. I startle a blue heron, and I can hear the beats of its wings as it sails off, affronted. I settle on the cold metal to drink my coffee and listen.

The story proceeds as the moon sets (at 7:36 A.M.) and as, behind me, imperceptibly, the sun rises (at 7:41 A.M.). The nodules, our narrator has heard on the news, are valuable for the rare-earth metals they contain and how they will power the rising volume of ". . . all the cars and turbines and machines." She is leased to the scientific company prospecting for the nodules, and prepares to dive down. I'm riveted. And I feel called out, particularly listening on a phone that runs on hardware made of rare-earth metals.

The yellow pilot boat has come to life and suddenly rushes past me. Behind it bubbles roil in a heave of whitewash. PACIFIC SPIRIT, says little block letters on her bow as she heads out to the vessels hulking in the fog, carrying or picking up the pilot who guides these vessels, slowly, through these rich and profoundly acoustic waters.

The moon is full now, below the horizon. My coffee is empty, the jar cold. I stand and wander farther down the breakwater.

EPILOGUE

Every three years, many members of the global bioacoustics community gather for a conference called "The Effects of Noise on Aquatic Life." The proceedings inform reports, policy, future research, and collaborations. In 2022, in a Berlin hotel, hundreds gather again. They carry tote bags stamped with the conference logo, a suite of marine animals from sketches by Hawkins.

Arthur Popper opens the conference via Zoom from Maryland, welcoming everyone. Tony Hawkins makes a short statement.

Joe Luczkovich, who worked with Rodney Rountree on fish sounds, points out to me in the lunch line that the fish in the chafing dish is a soniferous species. Then he explains how pressure and particle motion can become untethered in shallow water.

Someone mentions that when this conference first began, there was more focus on the impacts of seismic noise: Now, most of the sessions focus on pile driving. Louise Roberts, Tony Hawkins's student, gives a talk about vibrational communication. Joe Sisneros, the chair of the first morning's session, stands and exhorts everyone to please wear their masks when they're not eating and drinking. I pull aside Roberto Rocca of JASCO and ask him if he can tell me about Joe Scrimger, who worked with him at the time, discovering the sound of snow on water in Cowichan Bay.

Valeria Vergara is not here, but in the field, perched beside the St. Lawrence, atop a tower.

As we mill in the poster session, Francis Juanes gives me advice for making connections through Spanish airports on the way home.

Kieran Cox appears, looking dapper in a black dress shirt printed with artichokes, and typically energetic. He's excited to meet the people who have built the field, who have written the standards.

One morning, it is announced to a visibly emotional room that Sam Ridgway, the first dolphin doctor, passed away the evening before.

AT TIMES I have felt uncomfortable devoting so much of my life in the past two years to sound underwater. For all the wonders and worries of this subject, the truth is that noise does not match the deep threat posed to the oceans by climate change. And yet, neither issue is monolithic or exists in a vacuum. Warming or acidifying waters will conduct sound differently: Sound's effect on ecosystems like reefs or Arctic food webs will ricochet into animals' responses to climate change.

Yet I believe that it is never a waste to examine the world though a new lens, through a new sense. Underwater, sound behaves differently than in air. It travels farther, faster, and attenuates slower. You do not hear as an animal underwater would hear; you cannot. This fact alone I find both calming and exciting. Underwater animals show us how the thing we call an ear is only one way to perceive sound. The vast number of watery species tap into particle motion and not pressure— a completely different aspect of sound. This may seem a semantic difference. But imagine the ways that waves interfere in a wave pool, such as standing waves, and other phenomena, and you start to see (or, hear) the possibilities. What is that *like*? Also amazing to me is the emerging science on vibrational communication. It links touch and sound using structures such as lateral lines. And while we're at it, be amazed that those hair cells testify to the deep evolutionary history of detecting sound underwater. We've been doing this for a long, long time.

This science is accelerating, finally, as we start to understand, measure, and consider sound underwater where we didn't before. It's also a tempting thing to study for another reason: because noise seems like it should be one of the easier pollutants to fix.

But that's not always the case. Sure, "cleanup" is easier than an oil spill, say. But dizzy lobsters, bruised scallops, and confused mammals testify that noise harms can be subtle and long-lasting. We can't expect obvious answers or quick fixes. Noise is often a compounding problem in animals who are already stressed, hungry, or sick. And honestly, sometimes noise isn't a problem. Many animals have adapted or can adapt or aren't affected at all.

So, what should we do? Scientists are working to fill the data gaps on what sounds matter to invertebrates and how our noises and vibrations impact them. This means measuring the particle motion of a sound, not just its pressure, and looking at seabed vibrations, too.

Researchers are also using new technologies such as AI to process cumbersomely huge acoustic data files quickly. They are scrambling to get good data from areas like the Arctic and the deep sea that will enable us to learn more about these relatively pristine environments, using a multitude of frameworks: soundscapes, baseline monitoring, detailed old-school field surveys.

Non-acousticians like Kieran Cox are incorporating sound into their work, and bioacousticians continue to dig into what the complex vocalizations of marine mammals can tell us about their lives. They are building databases, developing new metrics, and even collaborating with artists to make the issue more accessible to the public. And in the near future, they will make some key decisions about landmark regulations.

Regulation is not a fix-all. A bad regulation is worse than no regulation creating resentment or corruption, and, coupled with the greenwashing that unfortunately pervades many conservation efforts, creates ample opportunity for failure. Regulation should be careful, simple, and effective.

Opinions on *how* to regulate may vary, but the science is getting clearer on *what* to regulate. Sonar, seismic, and pile driving can create dramatic impacts. But globally, noise-wise, shipping is the big one. Naval-architecture redesigns, slowdowns, and shipping-lane planning are useful. But at the end of the day, the culprit is our way of life: oil, consumer goods, even cruise ships. We cannot bemoan the noisy ocean and order all our household goods online.

Here's the thing: I don't think we needed the anthropause. The data was clear before Covid: Noise, especially loud, inescapable, constant, close, and incessant noise hurts marine life. None of the data I've seen emerge from the 2020 lockdowns has fundamentally changed this.

The science is also getting clearer on *who* regulations should protect. Marine mammals are critical but the lack of data on invertebrates, and their response to particle motion and seabed vibration, needs to be addressed.

And then there's the question of *where* and *when* regulations should take place. In addition to more populous or noisy waters, we have the chance to make good regulations before we move further into two regions for which we don't have a great deal of data: the Arctic, and the deep sea. We must do this soon.

As for *why*, well, there are good economic reasons to reduce noise, from preserving a valuable ocean to increasing ship efficiency. But for me, I confess, there is an intrinsic value in nature. Full stop. I think the world is so much bigger than we know. It's so much stranger, and so much richer, and asking about it, looking and listening to it, learning from it, makes us fully human. It's a priceless thing.

WHEN I RETURN from the Berlin conference, I visit our family home. I take the canoe out one evening on the lake, with the hydrophone, from the dock where my brother and I once played trucks. I dip the hydrophone gently into a weed bed and hear crunches, loud and clear, sounds I have never heard before, despite having grown up swimming these waters.

"It's probably the tiny little water boatmen," Krause tells me, later. These are insect larvae, chewing on the weeds as they grow.

Back on the dock it starts to rain. I drop the hydrophone and listen to the sound: strange, hissing like breath, unlike rain on any other surface. My brother, taking a break from family chaos upstairs, comes down to see what I'm doing, and I pass him the headphones.

ACKNOWLEDGMENTS

Too many people to name here have provided time, information, and assistance. Thank you to my editor, Libby Burton, and my extraordinary agent, Gillian MacKenzie, for believing in the project. To Heather, for the Sherwood chats and so much more. To Jude and Adrienne, my shoulder editors. To everyone who spoke with me for this project and trusted me with their work and stories, and those who read, reviewed, and checked the manuscript, especially: Erin Klenow, Spoorthy Raman, Sophie Weiler, Arthur Popper, and Steven Weinberg. These people have gone above and beyond to help me. Any remaining errors, omissions, or inaccuracies are entirely my own.

I want to thank several extraordinary teachers. Bob Roddie taught the Oceanography course where I first learned about coral and got excited about the ocean. Cynthia Rankin first told me I could be a writer and suggested I send my work out into the world. And the late Penny Park was one of the most impressive and generous people I have had the privilege of working with.

I am lucky to have many wonderful friends whose understanding, chats, and walks have kept me sane: Wendy, Kate, Allison, Sunita, Kari, Aaron, and the Victoria Group.

To my family: Libby, Neil, Blair, and Pupe. To Peter, for Spice Time. Especially to my grandmother, who passed away during this project, for always believing in me.

To Courtney, whose strength inspires me every day, and to Olivia: Welcome to a beautiful world. To my parents, who taught me everything I know and have supported me countless times in ways I'm just beginning to appreciate. To Wes, for the games of trucks and so much more.

Finally, to Warren: you've shown me what this life is for.

NOTES

INTRODUCTION

2   *The Silent World*: Jacques-Yves Cousteau and Louis Malle, dirs., *Le Monde du Silence* (The Silent World), 1956, 1:24:22 to 1:24:37.

2   During World War II: "Jacques Cousteau," *Encyclopedia Britannica*, August 11, 2023, https://www.britannica.com/biography/Jacques-Cousteau/.

2   "We have merely skimmed the surface": Cousteau, *Le Monde du Silence*.

3   through the nineteenth and twentieth centuries: J. B. Hersey, "A Chronicle of Man's Use of Ocean Acoustics," *Oceanus* 20, no. 2 (Spring 1977): 8–21. This article gives a lovely overview by a true insider, as Hersey was at the forefront of ocean acoustics and the development of seismic imaging in the mid-twentieth century.

3   social whales define their groups: Hal Whitehead and Luke Rendell, *The Cultural Lives of Whales and Dolphins* (Chicago: University of Chicago Press, 2015), 135, 209.

3   the fastest muscles in the animal kingdom: Michael L. Fine, Barbara Bernard, and Thomas M. Harris, "Functional Morphology of Toadfish Sonic Muscle Fibers: Relationship to Possible Fiber Division," *Canadian Journal of Zoology* 71, no 11 (November 1993): 2262, https://www.researchgate.net/publication/238031009_Functional_morphology_of_toadfish_sonic_muscle_fibers_Relationship_to_possible_fiber_division.

3   More recently, we've learned: Carlos M. Duarte et al., "The Soundscape of the Anthropocene Ocean," *Science* 371, no. 6529 (February 2021): 1 of 10, https://doi.org/10.1126/science.aba4658.

3 Even tiny larvae: S. D. Simpson et al., "Attraction of Settlement-Stage Coral Reefs Fishes to Ambient Reef Noise," *Marine Ecology Progress Series*, 276 (August 2004): 263.

3 Gear designed to sense earthquakes: Alexandre P. Plourde and Mladen R. Nedimovic, "Monitoring Fin and Blue Whales in the Lower St. Lawrence Seaway with Onshore Seismometers," *Remote Sensing in Ecology and Conservation* 8, no. 4 (August 2022): 551.

3 toothed-whale brethren: Whitehead and Rendell, *Cultural Lives*, 47.

3 unmatched by naval sonar: Haley Cohen Gilliland, "A Brief History of the US Navy's Dolphins," *MIT Technology Review*, October 24, 2019, https://www.technologyreview.com/2019/10/24/306/dolphin-echolocation-us-navy-war/.

4 "The movement of information": Whitehead and Rendell, *Cultural Lives*, 3.

4 Cousteau himself: Phillippe Cousteau and Patrick Watson, dirs., *Savage World of the Coral Jungle: The Undersea World of Jacques Cousteau*, YouTube video, 23:38–24:12, https://www.youtube.com/watch?v=iFjML77T6VY&ab_channel=Adventure-People/.

4 We hear ourselves in the sea: Duarte et al., "Soundscape."

5 "Noise" is a technical term: Anthony D. Hawkins and Arthur N. Popper, "A Sound Approach to Assessing the Impact of Underwater Noise on Marine Fishes and Invertebrates," *ICES Journal of Marine Science* 74, no. 3 (March–April 2017): 635–651, https://doi.org/10.1093/icesjms/fsw205/.

5 shipping noise in the ocean has doubled: Mark A. McDonald, John A. Hildebrand, and Sean M. Wiggins, "Increases in Deep Ocean Ambient Noise in the Northeast Pacific West of San Nicolas Island California," *The Journal of the Acoustical Society of America* 230, no. 120(2) (August 2006): 711–718, https://doi.org/10.1121/1.2216565/.

5 focused on marine mammals: United States Congress, "Marine Mammal Protection Act of 1972," U.S. Government Publishing Office, December 26, 2022, https://www.govinfo.gov/app/details/COMPS-1679.

5 fish and crabs: Arthur N. Popper and Anthony D. Hawkins, "The Importance of Particle Motion to Fishes and Invertebrates," *Journal of the Acoustical. Society of America* 143, no. 1 (January 2018): 470–488, https://doi.org/10.1121/1.5021594/.

5 scallops: Natacha Aguilar de Soto et al., "Anthropogenic Noise Causes Body Malformations and Delays Development in Marine Larvae," *Scientific Reports* 3, no. 2831 (October 2013), https://doi.org/10.1038/srep02831.

5 and even seagrass: Marta Sole et al., "Seagrass *Posidonia* Is Impaired by Human-Generated Noise," *Communications Biology* 4, no. 743 (2021), https://doi.org/10.1038/s42003-021-02165-3/.

6 fascinating research: Rodney A. Rountree and Francis Juanes, "Potential for Use of Passive Acoustic Monitoring of Piranhas in the Pacaya—Samiria National Reserve in Peru," *Freshwater Biology* 65, no. 1 (January 2020), 55–65, https://doi.org/10.1111/fwb.13185; J. Miguel Simoes et al., "Courtship and Agonistic Sounds by the Cichlid Fish Pseudotropheus Zebra," *Journal of the Acoustical Society of America* 124, no. 2 (January 2020), 1332–1338, https://doi.org/10.1121/1.2945712.

CHAPTER I

8 Bamfield Marine Sciences Centre: See https://bamfieldmsc.com/; the center is famous throughout this part of the world for housing leading marine research.

8 two dozen underwater research sites: Interview with Kieran Cox, September 2022.

8 large brown seaweeds: Maycira Costa et al., "Historical Distribution of Kelp Forests on the Coast of British Columbia: 1858–1956," *Applied Geography* 120 (June 2020), https://doi.org/10.1016/j.apgeog.2020.102230/.

9 more than a third of the world's coasts: Rochelle Baker, "The Ocean's Kelp Forest Are Worth Serious Coin: Report," *The Vancouver Sun*, April 21, 2023, online at https://vancouversun.com/news/local-news/the-oceans-kelp-forests-are-worth-serious-coin-report; see also Aaron M. Eger et al., "The Value of Ecosystem Services in Global Marine Kelp Forests," *Nature Communications* 14, no. 1894 (May 2023), https://doi.org/0.1038/s41467-023-37385-0/.

9 most of British Columbia's: Costa et al., "Historical Distribution."

9 sound is critical: G. Scowcroft et al., *Discovery of Sound in the Sea* (Kingston, RI: University of Rhode Island, 2018), 3.

9 more and more parts of the ocean: Carlos M. Duarte et al., "The Soundscape of the Anthropocene Ocean," *Science* 371, no. 6529 (February 2021), 1 of 10, https://doi.org/10.1126/science.aba4658/. This article offers some high-level examples. A more in-depth examination of the state of the research is found in the conference proceedings from the Effects of Noise on Aquatic Life conference. For example, at the most recent conference in Berlin, topics ranged from seagrass stems to drilling ships to squid hearing.

9   especially near coasts: Nathan D. Merchant et al., "Monitoring Ship
     Noise to Assess the Impact of Coastal Developments on Marine Mam-
     mals," *Marine Pollution Bulletin* 78 (2014): 85–95, https://doi.org/10
     .1016/j.marpolbul.2013.10.058/.

10   the math of scuba safety: Divers use a dive computer or a basic Rec-
     reational Dive Planner chart, such as the one seen here (https://www
     .a1scubadiving.com/wp-content/uploads/2018/06/PADI-Recreational
     -Dive-Table-Planner.pdf), which is based on time spent at a given depth
     and the necessary time needed to recover from that depth. Divers must
     also consider airplane flights, time since last dive, and other variables.

11   the shipping lane: "British Columbia Marine Transportation," (map)
     https://cmscontent.nrs.gov.bc.ca/geoscience/MapPlace1/Offshore
     /Coastal_Ports_Medium.pdf/.

11   a cluster of instruments: Christopher Barnes et al., "Understanding
     Earth–Ocean Processes using Real-Time Data from NEPTUNE, Can-
     ada's Widely Distributed Sensor Networks, Northeast Pacific," *Geosci-
     ence Canada*. 38 (2011): 21–30.

12   increases with vessel speed: Christine Erbe et al., "The Effects of Ship
     Noise on Marine Mammals—A Review," *Frontiers in Marine Science*
     6, no. 606 (October 2019): 3 of 21, https://doi.org/10.3389/fmars.2019
     .00606/.

13   in some places they *chorus*: See, for example, Craig A. Radford et al.,
     "Temporal Patterns of Ambient Noise of Biological Origin from a Shal-
     low Water Temperate Reef," *Oecologica*, 156, no. 4 (July 2008), 921–929,
     http://www.jstor.org/stable/40309581. However, scientists listening to
     the ocean were noting an "evening chorus" as early as the 1960s, for
     example, R. I. Tait, "The Evening Chorus: A Biological Noise Investiga-
     tion," Naval Research Laboratory, HMNZ Dockyard, Auckland (1962).

14   tiny phytoplankton: Rebecca Lindsey and Michon Scott, "What Are
     Phytoplankton?" NASA Earth Observatory, accessed September 11,
     2023, https://earthobservatory.nasa.gov/features/Phytoplankton/.

15   Water absorbs light more quickly: Scowcroft, *Discovery*, 3.

15   water carries chemicals: Scowcroft, *Discovery*, 3.

15   shivers with sound: Scowcroft, *Discovery*, 3; also Wenz, "Acoustic Am-
     bient Noise."

15   Urchins crunch kelp: Kelly Fretwell and Brian Starzomski, "Biodiversity
     of the Central Coast: Purple Sea Urchin," Central Coast Biodiversity,

accessed September 12, 2023, https://www.centralcoastbiodiversity.org
/purple-sea-urchin-bull-strongylocentrotus-purpuratus.html. The urchin's clicking and crunching was audible to me underwater.

15 clap their claws: Barbara Schmitz, "Sound Production in Crustacea
with Special Reference to the Alpheidae," in *The Crustacean Nervous System*, ed., K. Wiese (Berlin, Heidelberg: Springer, 2002), 1, https://doi
.org/10.1007/978-3-662-04843-6_40/.

15 Worms snap their jaws: Ryutaro Goto, Isao Hirabayashi, A. Richard
Palmer, "Remarkably Loud Snaps During Mouth-Fighting by a Sponge-
Dwelling Worm," *Current Biology* 29, no.13 (2019): 617–618, https://doi
.org/10.1016/j.cub.2019.05.047. This worm engages in a kind of "mouth
fighting," popping their jaws to make an extremely loud sound. For years
no one thought a soft-bodied worm could make such noise: the mechanism seems to be a very strong pharyngeal muscle.

15 Grunt sculpins: "Grunt Sculpin," Aquarium of the Pacific, accessed
September 11, 2023, https://www.aquariumofpacific.org/onlinelearning
center/species/grunt_sculpin/.

15 Black rockfish: "Sebastes melanops Girard, 1856: Black rockfish," Fish-
Base, accessed September 8, 2023, https://www.fishbase.se/summary
/sebastes-melanops.

18 began operating in 1972: "History," Bamfield Marine Science Centre,
accessed September 10, 2023, https://bamfieldmsc.com/bmsc-overview
/history/.

18 long "sounding lines": Susan Schlee, *The Edge of an Unfamiliar World*
(New York: E.P. Dutton & Co, 1973), 44.

19 in more detail than other animals: Most studies and regulation address
marine mammals, while the effects of sound on fish and invertebrates
is significantly less well studied, e.g., Hawkins and Popper, "Sound
Approach."

19 began mapping and surveying kelp: Summarized in, for example, Louis
D. Druehl, "The Distribution of *Macrocystis integrifolia* in British Co-
lumbia as Related to Environmental Parameters," *Canadian Journal of
Botany* 56 (1978): 69–79.

19 more have diminished: Samuel Starko et al., "Microclimate Predicts
Kelp Forest Extinction in the Face of Direct and Indirect Marine Heat-
wave Effects," *Ecological Applications* 32, no. 7 (2022), https://doi.org/10
.1002/eap.2673/.

20   several degrees Celsius by the year 2100: "Regional Fact Sheet: Ocean," IPCC, *Sixth Assessment Report*, https://www.ipcc.ch/report/ar6/wg1 /downloads/factsheets/IPCC_AR6_WGI_Regional_Fact_Sheet _Ocean.pdf/.

20   naturally about 2 degrees warmer: Interview with Kieran Cox, 2022.

20   an "ecosystem service": "Ecosystem Services," The National Wildlife Federation, accessed September 14, 2023, https://www.nwf.org/Educational -Resources/Wildlife-Guide/Understanding-Conservation/Ecosystem -Services/.

20   protecting coasts from storm surges: M. Spalding et al., "Mangroves for Coastal Defence. Guidelines for Coastal Managers & Policy Makers," Wetlands International and the Nature Conservancy (2014), 8, https://www.nature.org/media/oceansandcoasts/mangroves-for-coastal -defence.pdf/.

20   Forests clean the air: N. Balloffet et al., "Ecosystem Services and Climate Change," U.S. Department of Agriculture, Forest Service, Climate Change Resource Center (February 4, 2012), www.fs.usda.gov/ccrc/topics /ecosystem-services/.

20   Kelp-forest services are many: Aaron M. Eger et al., "The Value of Ecosystem Services in Global Marine Kelp Forests," *Nature Communications* 14, no. 1894 (May 2023), https://doi.org/0.1038/s41467-023-37385-0/.

21   updating its ship noise guidelines: "Addressing Underwater Noise From Ships—Draft Revised Guidelines Agreed," International Maritime Organization, January 30, 2023, https://www.imo.org/en/MediaCentre /Pages/WhatsNew-1818.aspx/.

21   cost-effective ship redesigns: Vanessa ZoBell et al., "Retrofit-Induced Changes in the Radiated Noise and Monopole Source Levels of Container Ships," *PLoS ONE* 18, no. 3 (2023), https://doi.org/10.1371 /journal.pone.0282677/.

21   guidelines for measuring underwater noise: International Organization for Standardization, "Standards by ISO/TC 43/SC 3, Underwater Acoustics," accessed September 2023, https://www.iso.org/committee /653046/x/catalogue/p/0/u/1/w/0/d/0/.

21   framework of concentric zones of influence: Hawkins and Popper, "Sound Approach."

22   only one organ: Sensory hairs, statocysts, and chordotonal organs are just some of the mechanisms besides an ear that seem to detect sound. See, for example, mussels: Louise Roberts et al., "Sensitivity of the Mussel

Mytilus Edulis to Substrate-Borne Vibration in Relation to Anthropogenically Generated Noise," *Marine Ecology Progress Series* 538 (2015): 185–195, https://doi.org/10.3354/meps11468. Also, there's work on sensory hairs of crayfish: J. Tautz and D. C. Sandeman, "The Detection of Waterborne Vibration by Sensory Hairs on the Chelae of the Crayfish," *Journal of Experimental Biology* 88, no. 1 (1980): 351–356, https://doi.org/10.1242/jeb.88.1.351. Longfin squid: T. A. Mooney et al., "Sound Detection by the Longfin Squid (*Loligo Pealeii*) Studied with Auditory Evoked Potentials: Sensitivity to Low-Frequency Particle Motion and Not Pressure," *Journal of Experimental Biology* 213, no. 21 (2010): 3748–3759, https://doi.org/10.1242/jeb.048348/.

22 movement of particles: Arthur D. Popper and Anthony D. Hawkins, "The Importance of Particle Motion to Fishes and Invertebrates," *Journal of the Acoustical Society of America* 143, no. 1 (January 2018): 470–488, https://doi.org/10.1121/1.5021594/.

22 sound is *different*: Robert J. Urick, *Principles of Underwater Sound, 3rd Edition* (McGraw-Hill, 1983): shallow water vs. deep, 209; bouncing and bending, 159; shadow zones, 135; channels and ducts, Chapter 6.

22 travels faster and farther: G. Scowcroft et al., *Discovery of Sound in the Sea* (Kingston, RI: University of Rhode Island, 2018), 5.

CHAPTER 2

23 Millions of species: "Great Barrier Reef," UNESCO, accessed September 2023, https://whc.unesco.org/en/list/154/; "Basic Information About Coral Reefs," United States Environmental Protection Agency, accessed September 2023, https://www.epa.gov/coral-reefs/basic-information -about-coral-reefs/.

23 releasing their respective gametes: "How Do Corals Reproduce?" NOAA, accessed September 2023, https://oceanservice.noaa.gov/education/tutorial _corals/coral06_reproduction.html/.

24 November new moon: S. D. Simpson et al., "Attraction of Settlement-Stage Coral Reef Fishes to Reef Noise," *Marine Ecology Progress Series* 276 (August 2004): 263–268, https://doi.org/10.3354/meps276263/.

24 had absolutely no idea: See G. P. Jones et al., "Self-Recruitment in a Coral Reef Fish Population," *Nature* 402 (1999): 802–804; S. E. Swearer et al., "Larval Retention and Recruitment in an Island Population of a Coral-Reef Fish," *Nature* 402 (1999): 799–802.

24   two studies radically changed his mind: Jones, "Self-Recruitment";
      Swearer, "Larval Retention."

24   Larval fish do have eyes: Michael J. Kingsford et al., "Sensory Environ-
      ments, Larval Abilities, and Local Self-Recruitment," *Bulletin of Marine
      Science*, Supplement 70, no.1 (2002): 309–340.

24   (recruitment did synchronize to moon phases): Simpson et al., "Attraction."

25   "There were reports": See, for example, Martin W. Johnson, "A Survey of
      Biological Underwater Noises Off the Coast of California and in Upper
      Puget Sound," University of California Division of War Research, U.S.
      Navy Radio and Sound Laboratory, San Diego, California (1943), which
      describes water *noise* surveys along the West Coast.

25   underwater listening networks: "Origins of SOSUS," COMSUBPAC,
      https://www.csp.navy.mil/cus/About-IUSS/Origins-of-SOSUS/.

25   finally got to hear: William A. Watkins and Mary Anne Daher, "Twelve
      Years of Tracking 52-Hz Whale Calls from a Unique Source in the
      North Pacific," *Deep Sea Research Part 1*, 51 (2004) 1889–1901, https://
      doi.org/10.1016/j.dsr.2004.08.006. "SOSUS provided convenient, accu-
      rate, and well-tested acoustic tracking, although over the next 40 years,
      such classified Navy data usually were not available for biological study.
      Then in 1992, data from US Navy Integrated Sound Surveillance Sys-
      tems including SOSUS were partially declassified . . ." Also, described in
      interview with Steve Simpson, 2022. Note that some acousticians such
      as William Watkins worked with SOSUS before its declassification, but
      many in the broader research community, such as Simpson, describe
      their knowledge of the network as low to nil.

25   Earth's seismic mumblings: Carlos M. Duarte et al., "The Soundscape
      of the Anthropocene Ocean," *Science* 371, no. 6529 (February 2021):
      1 of 10, https://doi.org/10.1126/science.aba4658; Gordon M. Wenz,
      "Acoustic Ambient Noise in the Ocean: Spectra and Sources," *Journal of
      the Acoustical Society of America* 34, no. 12 (December 1962): 1936–1956.

25   an unexpected thunder: "Underwater Microphone Captures Honshu,
      Japan Earthquake," NOAA PMEL, uploaded to YouTube April 12, 2011,
      https://www.youtube.com/watch?v=4rWDrZIucAQ&ab_channel
      =NOAAPMEL/.

25   Wind and wave sound: Wenz, "Acoustic Ambient Noise."

25   Heavy rain: J. A. Nystuen and D. M. Farmer, "The Influence of Wind on
      the Underwater Sound Generated by Light Rain," *Journal of the Acoustical
      Society of America* 82 (1987): 270–274, https://doi.org/10.1121/1.395563/.

25 Even falling snow makes a sound: J. Scrimger, "Underwater Noise Caused by Precipitation," *Nature* 318 (1985): 647–649, https://doi.org /10.1038/318647a0/.

26 haunting, echoing: This is my interpretation of the sound that I hear on any deep-water recording.

26 sea ice creaks and booms: Interview with William Halliday, 2021.

26 famously dubbed the "Bloop": "What Is the Bloop?" NOAA, National Ocean Service website, last updated January 18, 2023, https:// oceanservice.noaa.gov/facts/bloop.html/.

26 low-pitched sounds: R. P. Dziak et al., "Life and Death Sounds of Iceberg A53a," *Oceanography* 26, no. 2 (2013): 10–12, https://doi.org/10 .5670/oceanog.2013.20.

26 "chorusing": Douglas H. Cato, "Marine Biological Choruses Observed in Tropical Waters Near Australia," *Journal of the Acoustical Society of America* 64 (1978): 736–743, https://doi.org/10.1121/1.382038/.

27 very good sense for warning: Seth Horowitz, *The Universal Sense* (New York: Bloomsbury, 2012), 108–109; T. Götz and V. M. Janik, "Repeated Elicitation of the Acoustic Startle Reflex Leads to Sensitisation in Subsequent Avoidance Behaviour and Induces Fear Conditioning," *BMC Neuroscience* 12, no. 30 (2011), https://doi.org/10.1186/1471-2202-12-30. The latter states that the mammalian startle response is triggered by touch, gravity, or sound, of which only sound is a "distance" sense. It's probably not a coincidence that, as we'll see, touch, gravity, and sound as senses are quite closely linked.

27 tracked larval fish: Jeffrey M. Leis, Brooke M. Carson-Ewart, and Douglas H. Cato, "Sound Detection In Situ by the Larvae of a Coral-Reef Damselfish (Pomacentridae)," *Marine Ecology Progress Series* 232 (2002): 259–268. Note the data was collected beginning in 1998.

27 attracted baby fish: N. Tolimeri, A. Jeffs, and J. C. Montgomery, "Ambient Sound As a Cue for Navigation by Pelagic Larvae of Reef Fishes," *Marine Ecology Progress Series* 207 (2000): 219–224.

27 "Homeward Sound": Stephen D. Simpson et al., "Homeward Sound," *Science* 308 (April 2005): 221, https://doi.org/10.1126/science.1107406/.

28 "they haven't got a brain": Interview with Steve Simpson, 2022. The basic biology of a coral doesn't include ears.

28 just enough to sink or rise: Interview with Steve Simpson, 2022; Mark J. A. Vermeij et al., "Coral Larvae Move Toward Reef Sounds," *PLOSOne* 5, no. 5 (2010), https://doi.org/10.1371/ journal.pone.0010660/.

28   moved toward the sound: Vermeij, "Coral Larvae Move Toward Reef Sounds."

28   metamorphose into adults: Jenni A. Stanley et al., "Induction of Settlement in Crab Megalopae by Ambient Underwater Reef Sound," *Behavioral Ecology* 21, no. 1 (January–February 2010): 113–120, https://doi.org/10.1093/beheco/arp159/.

29   earliest simple cells: Peter Godfrey-Smith, *Other Minds* (New York: Farrar, Strauss and Giroux, 2016), 15. Note that the most common figure is actually 3.8 billion years ago, which means by three and a half billion years ago, they were almost certainly wobbling.

29   first "senses": Godfrey-Smith, *Other Minds*, 17.

29   2 billion years before: Ben Warren and Manuela Nowotny, "Bridging the Gap Between Mammal and Insect Ears: A Comparative and Evolutionary View of Sound-Reception," *Frontiers in Ecology and Evolution* 9 (2021), https://doi.org/10.3389/fevo.2021.667218/.

29   different sense organs: Godfrey-Smith, *Other Minds*, 18; Warren and Nowotny, "Bridging the Gap."

29   didn't have hard parts: T. Fridtjof Flannery, "Cambrian Explosion," *Encyclopedia Britannica*, accessed September 12, 2023, https://www.britannica.com/science/Cambrian-explosion.

29   moved with structures such as cilia: Warren and Nowotny, "Bridging the Gap."

29   the Cambrian Explosion: Warren and Nowotny, "Bridging the Gap."

30   six kingdoms: T. Cavalier-Smith, "Only six kingdoms of life," *Proceedings of the Royal Society of London. Series B: Biological Sciences* 271 (2004): 1251–1262.

30   phylum Chordata: Note that, strictly speaking, all vertebrates are chordates, but not all chordates have a backbone, the defining feature of a vertebrate. However, the term "vertebrate" is usually used broadly for chordates.

30   differences between the genes: Interview with Allison Coffin, 2022.

30   most animals on the planet: "What Is an Invertebrate?" Exploring Our Fluid Earth, University of Hawaii at Manoa, accessed September 2023, https://manoa.hawaii.edu/exploringourfluidearth/biological/invertebrates/what-invertebrate; also the IUCN's species list, "IUCN Red List," version 2014.3, Table 1, last updated November 13, 2014, http://cmsdocs.s3.amazonaws.com/summarystats/2014_3_Summary_Stats_Page_Documents/2014_3_RL_Stats_Table_1.pdf/.

30   Approximately 90 percent: Emily Yi-Shuyuan Chen, "Often Over-
      looked: Understanding and Meeting the Current Challenges of Marine
      Invertebrate Conservation," *Frontiers of Marine Science* 8, https://doi.org
      /10.3389/fmars.2021.690704/.

31   most phyla had specialized structures: Allison Coffin et al., "Evolution
      of Sensory Hair Cells," in *Evolution of the Vertebrate Auditory System*, eds.
      Geoffrey A. Manley, Arthur N. Popper, and Richard A. Fay (New York:
      Springer, 2004), 55–94, especially 58.

31   when these hair cells evolved: Jeremy S. Duncan and Bernd Fritzsch,
      "Evolution of Sound and Balance Perception: Innovations that Aggre-
      gate Single Hair Cells into the Ear and Transform a Gravistatic Sensor
      into the Organ of Corti," *The Anatomical Record* 295, no. 11 (November
      2022): 1760–1774, https://doi.org/10.1002/ar.22573/.

31   cnidarians: Ethan Ozment et al., "Cnidarian Hair Cell Development
      Illuminates an Ancient Role for the Class IV POU Transcription Fac-
      tor in Defining Mechanoreceptor Identity," *eLife* 10 (2021): e74336,
      https://doi.org/10.7554/eLife.74336/.

31   statocysts: Coffin, "Sensory Hair Cells," 79.

31   sense some sounds: R. J. Pumphrey, "Hearing," *Symposia for the Society of
      Experimental Biology* 4 (1950): 3–18.

32   a pressure wave: interview with Jim Miller, 2021: "it's a pressure wave."
      Also David L. Bradley and Richard Stern, "Underwater Sound and the
      Marine Mammal Acoustic Environment: A Guide to Fundamental Prin-
      ciples," prepared for the U.S. Marine Mammal Commission, 2008, page 1.

32   "disturbance or variation": *Merriam-Webster Dictionary*, s.v. "wave," ac-
      cessed September 15, 2023, https://www.merriam-webster.com/dictionary
      /wave.

32   *longitudinal* wave: "What Is Sound?" DOSITS, accessed September 15,
      https://dosits.org/science/sound/what-is-sound/.

32   *amplitude*: "Amplitude," DOSITS, accessed September 15, https://dosits
      .org/glossary/amplitude/.

33   Frequency: "Frequency," DOSITS, accessed September 15, https://dosits
      .org/glossary/frequency/; Bradley and Stern, "Underwater Sound," 7.

33   human with good ears: "The Audible Spectrum," in *Neuroscience, 2nd Edi-
      tion*, eds. D. Purves et al. (Sunderland, MA: Sinauer Associates; 2001).
      Available from https://www.ncbi.nlm.nih.gov/books/NBK10924/.

33   density very close to water: Manley, Popper, and Fay, *Evolution of the
      Vertebrate Auditory System*, x.

34 "It is impossible": Pumphrey, "Hearing," 14.
34 abdominal sense organ: Jeroen Hubert et al., "Responsiveness and Habituation to Repeated Sound Exposures and Pulse Trains in Blue Mussels," *Journal of Experimental Marine Biology and Ecology* 547 (February 2022), https://doi.org/10.1016/j.jembe.2021.151668/.
34 hair-cell-filled pockets: Interview with Jayson Semmens, 2021.
34 structures in their joints: Arthur N. Popper, Michael Salmon, and K. W. Horch, "Acoustic Detection and Communication by Decapod Crustaceans," *Journal of Comparative Physiology A-Neuroethology Sensory Neural and Behavioral Physiology* 187 (2021): 83–89.
34 Modern invertebrates often respond best: Pumphrey, "Hearing," 10–11.
34 "touch at a distance": Sven Dijkgraaf, "Spallanzani's Unpublished Experiments on the Sensory Basis of Object Perception in Bats," *Isis* 51, no. 9–20 (March 1960), https://doi.org/10.1086/348834; PMID: 13816753.
35 cilia on their bodies: Vermeij, "Coral Larvae Move Toward Reef Sounds."
35 lateral line: Sheryl Coombs et al., eds., *The Springer Handbook of Auditory Research: The Lateral Line* (New York: Springer, 2014), Preface.
36 inside ancestral fishes' heads: Fred Ladich and Arthur N. Popper, "Parallel Evolution in Fish Hearing Organs," in Manley, Popper, and Fay, *Evolution of the Vertebrate Auditory System*, 97.
36 the fish's inner ear: Geoffrey A. Manley, Arthur N. Popper, Richard A. Fay, eds., *Evolution of the Vertebrate Auditory System* (New York: Springer, 2004), x.
37 use these stones for valuable data: Interview with Luke Tornabene and Katherine Maslenikov; "Two Million Fish Ear Bones Contain New Environments Insights," Burke Museum, updated October 24, 2012, https://www.burkemuseum.org/news/two-million-fish-ear-bones-contain-new-environmental-insights/.
38 have long argued: "The Importance of Sound to Fishes—Arthur Popper and Tony Hawkins | ABC 2020," Sea Search Research and Conservation, https://www.youtube.com/watch?v=hJGuk4JVXuU&ab_channel=SeaSearchResearch%26Conservation/.
38 air-filled sac: Tanja Schulz-Mirbach, Brian Metscher, and Friedrich Ladich, "Relationship Between Swim Bladder Morphology and Hearing Abilities: A Case Study on Asian and African Cichlids," *PLOSOne* 7, no. 8 (August 7, 2012), https://doi.org/10.1371/journal.pone.0042292/.
38 "hearing specialists": Arthur N. Popper, Anthony D. Hawkins, and

Joseph A. Sisneros, "Fish Hearing 'Specialization'—A Re-Evaluation," *Hearing Research* 425 (2022), https://doi.org/10.1016/j.heares.2021 .108393; interview with Arthur Popper; Friedrich Ladich and Lidia Eva Wysocki, "How Does Tripus Extirpation Affect Auditory Sensitivity in Goldfish?" *Hearing Research* 182, no. 1–2 (2003): 119–129, https://doi .org/10.1016/S0378-5955(03)00188-6.

39   400 million years ago: Gemma Tarlach, "How Life First Left Water and Walked Ashore," *Discover Magazine,* June 12, 2017, https://www.discovermagazine.com/planet-earth/how-life-first-left -water-and-walked-ashore/.

39   "The chemistry of life is an aquatic chemistry": Godfrey-Smith, *Other Minds,* 19.

39   *impedance*: Interview with Jim Miller, 2021.

39   100 million years: Geoffrey A. Manley, "Advances and Perspectives in the Study of the Evolution of the Vertebrate Auditory System," in Manley, Popper, and Fay, *Evolution,* 366.

40   an amplifying chain of small bones: Jennifer A. Clack and Edward Allin, "The Evolution of Single- and Multiple-Ossicle Ears in Fishes and Tetrapods," in Manley, Popper, and Fay, *Evolution,* 146.

40   Amphibians: Geoffrey A. Manley and Jennifer Clack, "An Outline of the Evolution of Vertebrate Hearing Organs," in Manley, Popper, and Fay, *Evolution,* 1-4. Note that the line wasn't direct.

41   Divers wearing full neoprene hoods: D. M. Fothergill, J. R. Sims, and M. D. Curley, "Neoprene Wet-Suit Hood Affects Low-Frequency Underwater Hearing Thresholds," *Aviation, Space, and Environmental Medicine* 75, no. 5 (May 2004): 397-404, PMID: 15152891.

CHAPTER 3

43   the Greek philosopher Aristotle: Christopher Shields, "Aristotle," in *The Stanford Encyclopedia of Philosophy* (Spring 2022 Edition), ed. Edward N. Zalta, https://plato.stanford.edu/archives/spr2022/entries/aristotle/.

43   "sound is heard": Aristotle, *De Anima, Part 2, Book 8,* trans. J. A. Smith, MIT Classics, http://classics.mit.edu/Aristotle/soul.2.ii.html.

43   "the lyra or gurnard": Aristotle, *Historium Animalium Book 4, Part 9,* trans. D'Arcy Wentworth Thompson (350 BCE), The Internet Classic Archive, http://classics.mit.edu/Aristotle/history_anim.mb.txt/.

44   "internal parts about their bellies": Aristotle, *Historium Animalium.*

44   "Now, the seal": Aristotle, *Historium Animalium, Book 4, Part 11.*

44   *piscinae*: James Arnold Higginbotham, *Piscinae: Artificial Fishponds in Roman Italy* (Chapel Hill, NC: University of North Carolina Press, 1997).

44   "In the Emperor's aquarium": Pliny the Elder, *Natural History, Volume III:* Books 8–11. Translated by H. Rackham. Loeb Classical Library 353: 415-416, https://www.loebclassics.com/view/pliny_elder -natural_history/1938/pb_LCL353.415.xml?readMode=recto.

44   Marcus Licinius Crassus: "Marcus Licinius Crassus," *Encyclopedia Britannica* (August 29, 2023), https://www.britannica.com/biography/Marcus -Licinius-Crassus.

44   adorned her eel with jewelry: Higginbothem, Higginbotham, *Piscinae,* 63.

44   Leonardo da Vinci mused: Robert J. Urick, *Principles of Underwater Sound, 3rd Edition* (New York: McGraw-Hill, 1983), 2.

44   the anatomist Bartholomeus Eustachius: A. H. Gitter, "Eine kurze Geschichte der Hörforschung" (A short history of hearing research), *Laryngorhinootologie* 69, no. 9 (September 1990): 495–500, https://doi.org /10.1055/s-2007-998239, PMID: 2242190; T. R. De Water, "Historical Aspects of Inner Ear Anatomy and Biology that Underlie the Design of Hearing and Balance Prosthetic Devices," *The Anatomical Record,* 295 (2012): 1741–1759, https://doi.org/10.1002/ar.22598/.

45   lavishly illustrated: Giulio Casserio, "De vocis auditusque organis historia anatomica, singulari fide, methodo ac industria concinnata, tractatibus duobus explicata ac variis iconibus aere excusis illustrata," https://gallica.bnf.fr/ark:/12148/bpt6k850342w/f6.item; see, for example, https://commons.wikimedia.org/wiki/File:Casserius,_De_Vocis _Auditusque_Organis_Historia_Anatomica._Wellcome_L0007973 .jpg/.

45   Copernicus' declaration: Stephen Hawking, "Nicolaus Copernicus (1473–1543) His Life and Work," in *On the Shoulders of Giants,* ed. Stephen Hawking (Philadelphia: Running Press, 2002), 1–6.

45   *Novum Organum*: Francis Bacon, *The New Organon* (New York: Jonathan Bennett, 2017), https://www.earlymoderntexts.com/assets/pdfs /bacon1620.pdf.

45   (Bacon also made mention): Francis Bacon, *The Works of Francis Bacon, Baron of Verulam, Viscount St. Alban, and Lord High Chancellor of England, in Five Volumes* (A. Millar in the Strand, 1765), 296–297.

45   It's hard to overstate: Hawking, "Nicolaus Copernicus," 1–6.

45   Linnaeus created: "Carolus Linnaeus summary," *Encyclopedia Britannica*, May 2, 2020, https://www.britannica.com/summary/Carolus-Linnaeus.

45   *On the Origin of Species*: Charles Darwin and Leonard Kebler, *On the origin of species by means of natural selection, or, The preservation of favoured races in the struggle for life* (London: J. Murray, 1859), https://www.loc.gov/item/06017473/.

46   But fish were puzzling: The debate is described in various accounts, including John Hunter, "Account of the Organ of Hearing in Fish. By John Hunter, Esq. F. R. S," *Philosophical Transactions of the Royal Society of London* 72 (1782): 379–83, http://www.jstor.org/stable/106467; https://royalsocietypublishing.org/doi/pdf/10.1098/rstl.1782.0025/.

46   *The Compleat Angler*: Isaak Walton, *The Complete Angler,* cited in G. H. Parker, "A Critical Survey of the Sense of Hearing in Fishes," *Proceedings of the American Philosophical Society* 57, no. 2 (1918): 69–98.

46   strolling about a fishpond: Hunter, "Organ of Hearing in Fish," 379–383.

46   one shotgun blast: Parker, "A Critical Survey," 69–98.

47   German ethologist Karl von Frisch: Karl von Frisch, *A Biologist Remembers,* trans. Lisbeth Gombrich (New York: Pergamon, 1967).

48   Naturalists guessed: Roger Hanlon, Roger Budelmann, and Bernd Budelmann, "Why Cephalopods Are Probably Not 'Deaf,'" *American Naturalist* 129 (1987), https://doi.org/10.1086/284637.

49   Marquis de Laplace: Bernard S. Finn, "Laplace and the Speed of Sound," *ISIS* 55, no. 179 (1964): 7–19, https://www3.nd.edu/~powers/ame.20231/finn1964.pdf/.

49   Jean-François Nollet: L'Abbe Nollet, "Sur l'Ouïe des Poissons, & sur la transmission des sons dans l'eau," *Histoire de l'Académie royale des sciences* (1743), http://visualiseur.bnf.fr/CadresFenetre?O=NUMM-3541&M=tdm/.

49   Alexander Monro Secundus: J. B. Hersey, "A Chronicle of Man's Use of Ocean Acoustics," *Oceanus* 20, no. 2 (Spring 1977): 8–21.

50   François Sulpice Beudant: F. S. Beudant, *Essai d'un Cours Elementaire et General des Sciences Physiques: Partie Physique* (Verdiere, Paris, 1824). Also Hersey says that they reported Beudant's figures in their own results: Hersey, "Man's Use of Ocean Acoustics."

50   designed an experiment: Stéphane Fischer, "Jean-Daniel Colladon, Geneva Scholar and Industrialist," Musée d'histoire des sciences, Ville de

Genève, http://institutions.ville-geneve.ch/fileadmin/user_upload/mhn /documents/Musee_histoire_des_sciences/8_Colladon_Angl_2.pdf/.

51   reported the speed of sound in water: J. D. Colladon and C. F. Sturm, "Memoire sur la compression des liquides: I. Introduction," *Annales de chimie.et de physique* 36 (1827): 113–159.

51   bumps through air: "Speed of Sound," Simon Fraser University, accessed September 14, 2023, https://www.sfu.ca/sonic-studio-webdav /handbook/Speed_Of_Sound.html/.

51   loses energy less quickly: G. Scowcroft et al., *Discovery of Sound in the Sea* (Kingston, RI: University of Rhode Island, 2018), 3.

51   "to use the echo": Fischer, "Colladon."

51   commerce was booming: Susan Schlee, *The Edge of an Unfamiliar World* (New York: E.P. Dutton & Co, 1973), 13–15.

51   Captain Cook: Rebecca J. Rosen, "For Scientists of the 18th Century, the Transit of Venus Was Their Final Chance to Measure the Solar System," *The Atlantic*, June 5, 2012, https://www.theatlantic.com/technology /archive/2012/06/for-scientists-of-the-18th-century-the-transit-of -venus-was-their-final-chance-to-measure-the-solar-system/258013/.

52   they needed maps: Schlee, *Unfamiliar World*, 23.

52   U.S. Coast Survey: Schlee, *Unfamiliar World*, 24.

52   Depot of Charts and Instruments: Schlee, *Unfamiliar World*, 26.

52   they had only sounding lines: Schlee, *Unfamiliar World*, 43–44.

52   During one three-and-a-half-year voyage: Schlee, *Unfamiliar World*, 251.

52   "an audible echo": "Proceedings of the American Philosophical Society, Progress of Physical Science," *Journal of the Franklin Institute of the State of Pennsylvania and Mechanics' Register* 24 (July 1839): 351–352.

53   repeated Bonnycastle and Patterson's experiment: "The First Studies of Underwater Acoustics: The 1800s," DOSITS, accessed September 13, 2023, https://dosits.org/people-and-sound/history-of-underwater -acoustics/the-first-studies-of-underwater-acoustics-the-1800s/.

53   The first subsea telegraph cables: Schlee, *Unfamiliar World*, 89.

53   Engineer Arthur Mundy: Transmission of Sound, United States Patent Office, No. 636519, Filed April 14, 1899, issued November 7, 1899, accessed September 12, 2023, https://patents.google.com/patent/US63 6519A/en.

53   Submarine Signal Company: Hersey, "Man's Use of Ocean Acoustics," 11

53   submerged in water tanks: C. Borbach, "An Interlude in Navigation: Submarine Signaling as a Sonic Geomedia Infrastructure," *New*

Media & Society 24, no. 11 (2022): 2493–2513, https://doi.org/10.1177 /14614448221122240/.

53 a chilly April day in 1914: R. F. Blake, "Submarine Signaling: The Protection of Shipping by a Wall of Sound and Other Uses of the Submarine Telegraph Oscillator," *Transactions of the American Institute of Electrical Engineers Vol. XXXIII, Part II: 1549–1561* (American Institute of Electrical Engineers, 1914).

54 Sir Hiram Maxim: "Sir Hiram Maxim's Plan to Prevent Sea Collisions," *New York Times*, July 28, 1912, https://timesmachine.nytimes .com/timesmachine/1912/07/28/100589415.html?pageNumber=45/.

54 Fessenden built a prototype: Blake, "Submarine Signaling."

54 Fessenden was unable to sleep: Helen Fessenden, *Fessenden: Builder of Tomorrows* (New York: Coward-McCann, 1940), 220.

54 another echo: Blake, "Submarine Signaling."

55 laid up with tuberculosis: Schlee, *Unfamiliar World*, 247–249; Urick, *Principles*, 3.

55 British ships were outfitted with hydrophones: Schlee, *Unfamiliar World*, 248.

55 A higher-frequency sound: Interview with Jim Miller, 2021.

55 existing oscillators: Angela D'Amico and Richard Pittinger, "A Brief History of Active Sonar," *Aquatic Mammals* 35, no. 4 (2009): 426–434, see Table 1.

55 A German U-boat of the time was about 60 meters: "Germans Unleash U-Boats," history.com, last updated January 28, 2021, https://www.history .com/this-day-in-history/germans-unleash-u-boats. These dimensions were for World War I; future submarine dimensions would obviously change.

56 French physicist Paul Langevin: Francis Duck, "Paul Langevin, U-boats, and Ultrasonics," *Physics Today* 175, no. 11 (November 2022): 42–48, https://doi.org/10.1063/PT.3.5122.

56 a warehouse of quartz: Duck, "Langevin."

56 send a signal 8 kilometers: David Zimmerman, "Paul Langevin and the Discovery of Active Sonar or Asdic," *The Northern Mariner/Le marin du nord* 12, no. 1 (January 2002): 39–52, https://www.cnrs-scrn.org /northern_mariner/vol12/tnm_12_1_39-52.pdf/.

56 "ASDICS": Robert W. Morse, "Acoustics and Submarine Warfare," *Oceanus* 20, no. 2 (Spring 1977): 69.

57 the SSC sold an echo-sounding "fathometer": Hersey, "Man's Use of Ocean Acoustics," 13.

57   the first submarine cables: Schlee, *Unfamiliar World*, 250–251.

57   "afternoon problem": Columbus O'Donnell Iselin and A. H. Woodcock, "Preliminary Report on the Prediction of 'Afternoon Effect,'" Woods Hole Oceanographic Institution, July 25, 1942, https://darchive .mblwhoilibrary.org/entities/publication/0866ee5b-afe1-5161-857b -d92b7565f2f0/.

57   the chief of naval operations: (Archival documents from Woods Hole)

57   it worked differently in the morning: Hersey, "Man's Use of Ocean Acoustics," 13–14.

58   the sound may also *refract*: Interview with Jim Miller, 2021; "How Does sound Move? Refraction," DOSITS, accessed September 14, 2023, https://dosits.org/science/movement/how-does-sound-move/refraction/.

58   Three factors can affect seawater density: "Sound Transmission in the Ocean," *The Water Encyclopedia*, accessed September 13, 2023, http:// www.waterencyclopedia.com/Re-St/Sound-Transmission-in-the -Ocean.html/.

58   ray tracings: Urick, *Principles*, 159–160.

59   In a deep mid-latitude ocean: W. Munk, P. Worcester, and C. Wunsch, *Ocean Acoustic Tomography* (Cambridge, UK: Cambridge University Press, 1995). Note that the tropical latitudes reach a minimum speed at about 1 kilometer: the temperate ones, much higher.

59   In the Arctic: Urick, *Principles*, 169.

59   even thousands, of kilometers: Maurice W. Ewing and J. Lamar Worzel, "Long Range Sound Transmission: Interim Report No. 1, March 1, 1944—January 20, 1945," Woods Hole Oceanographic Institution, https://darchive.mblwhoilibrary.org/entities/publication/976cacef-a9ef -5db4-b3b4-e15232e4a267/.

60   In 1945, he tested his theory: Hersey, "Man's Use of Ocean Acoustics," 14.

60   SOFAR channel: Urick, *Principles*, 169.

60   In 1946: Gary E. Weir, "The American Sound Surveillance System: Using the Ocean to Hunt Soviet Submarines, 1950–1961," *International Journal of Naval History* 5, no. 2 (August 2006); https://www.usni.org /magazines/naval-history-magazine/2021/february/66-years-undersea -surveillance/.

60   to listen for enemy submarines: Donald P. Loye and Don A. Proudfoot, "Underwater Noise Due to Marine Life," *Journal of the Acoustical Society of America* 18 (1946), 446–449, https://doi.org/10.1121/1.1916386;

Winthrop Kellogg, *Porpoises and Sonar* (Chicago: The University of Chicago Press, 1961), 33–34.

61 "carpenter noises": Sam H. Ridgway, "Revealing the 'Carpenter Fish' and Setting the Hook for Bioacoustics in the U.S. Navy: Personal Reflections on Schevill and Watkins," in *Voices of Marine Mammals: William E. Schevill and William A. Watkins: Pioneers in Bioacoustics*, ed. Christina Connett Brophy (New Bedford, MA: The New Bedford Whaling Museum/Old Dartmouth Historical Society, 2019), 83–89.

61 20-Hz pulse: B. Patterson and G. R. Hamilton, "Repetitive 20 Cycle Per Second Biological Hydroacoustic Signals at Bermuda," in *Marine Bioacoustics, vol. 1,* ed. W. N. Tavolga (Oxford: Pergamon Press, 1964), 225–245.

62 a crackling: see, for example, Martin Johnson, "A Survey of Biological Underwater Noises Off the Coast of California and in Upper Puget Sound," *UCDWR* No. U100 (September 10, 1943).

62 "false bottom": David Levin, "The Mysterious False Bottom of the Twilight Zone," Woods Hole Oceanographic Institution, April 26, 2022, https://twilightzone.whoi.edu/the-mysterious-false-bottom-of-the-twilight-zone/.

62 Martin Johnson: Schlee, *Unfamiliar World,* 299–301.

62 "Snapping shrimp": Michel Versluis et al., "How Snapping Shrimp Snap: Through Cavitating Bubbles," *Science* 289 (2000): 2114–2117, https://doi.org/10.1126/science.289.5487.2114/.

63 "sea has long been looked upon": Martin W. Johnson, F. Alton Everest, and Robert W. Young, "The Role of Snapping Shrimp (Crangon and Synalpheus) in the Production of Underwater Noise in the Sea," *Biological Bulletin* 93, no. 2 (1947): 122–138, https://doi.org/10.2307/1538284.

63 William Schevill: Christina Connett Brophy, "Introduction," in *The Voices of Marine Mammals: William E. Schevill and William A. Watkins: Pioneers in Bioacoustics* (New Bedford, MA: The New Bedford Whaling Museum, 2019), 8.

63 Barbara Lawrence: Maria Rutzmoser, "Obituary: Barbara Lawrence Schevill: 1909–1997," *Journal of Mammalogy* 80, no. 3 (August 1999).

63 travel where you know: William E. Schevill and Barbara Lawrence, "A Phonograph Record of the Underwater Calls of Delphinapterus Leucas," Woods Hole Oceanographic Institution, January 1950.

64 genetically distinct population has remained: Fisheries and Oceans

Canada (DFO), "Recovery Strategy for the Beluga (Delphinapterus Leucas) St. Lawrence Estuary Population in Canada" [Proposed], Species at Risk Act, Recovery Strategy Series, Fisheries and Oceans Canada, Ottawa, 2011.

64   Two Tadoussac locals: William E. Schevill Barbara Lawrence, "Underwater Listening to the White Porpoise (Delphinapterus leucas)," *Science* 109 (1949): 143–144, https://doi.org/10.1126/science.109.2824.143/.

64   Schevill had some clues: L. Worthington and W. Schevill, "Underwater Sounds Heard from Sperm Whales," *Nature* 180 (1957): 291, https://doi.org/10.1038/180291a0/.

65   that low double-time blip: W. E. Schevill, W. A. Watkins, and R. H. Backus, "The 20-Cycle Signals and Balaenoptera (Fin whales)," in *Marine Bio-Acoustics: Proceedings of a Symposium Held at the Lerner Marine Laboratory, Bimini, Bahamas, April 11–13, 1963,* ed. William N. Tavolga (London: Pergamon Press, 1964), 147–152.

65   Schevill died in 2004: W. D. Ian Rolfe, "William Edward Schevill: Palaeontologist, Librarian, Cetologist, Biologist," *Archives of Natural History,* 39, no. 1 (April 2012): 162–164.

65   Marie Poland Fish: Ben Goldfarb, "Biologist Marie Fish Catalogued the Sounds of the Ocean for the World to Hear," *Smithsonian Magazine,* April 2021, https://www.smithsonianmag.com/science-nature/biologist-marie-fish-catalogued-sounds-ocean-world-hear-180977152/.

65   cattle prods: William N. Tavolga, "Fish Bioacoustics: A Personal History," *Bioacoustics* 12, no. 3 (2002): 101–104, https://doi.org/10.1080/09524622.2002.9753662/.

65   recorded some sounds in water: Marie Poland Fish and William H. Mowbray, *Sounds of Western North Atlantic Fishes, A Reference File of Biological Underwater Sounds* (Baltimore: The Johns Hopkins Press, 1970).

66   widest variety of sound-making structures: Friedrich Ladich et al., "Sound-Generating Mechanisms in Fishes: A Unique Diversity in Vertebrates" in F. Ladich, S.P. Collin, P. Moller, and B.G. Kapoor, eds: *Communication in Fishes,* Vol. 1 (Enfield, NH: Science Publishers, 2006): 3–43.

66   Sculpins move their: Michael Fine and Eric Parmentier, "Mechanisms of Fish Sound Production" in *Sound Communication in Fishes,* ed. F. Ladich (New York: Springer, 2015), 10.1007/978-3-7091-1846-7_3.

66   Toadfish: Michael L. Fine, "Sexual Dimorphism of the Growth Rate

of the Swimbladder of the Toadfish Opsanus Tau," *Copeia*, no. 3 (1975): 483–490, https://doi.org/10.2307/1443646.

66   squirrelfish: Howard E. Winn and Joseph A. Marshall, "Sound-Producing Organ of the Squirrelfish, Holocentrus Rufus," *Physiological Zoology* 36, no. 1 (1963): 34–44, http://www.jstor.org/stable/30152736.

66   Herring are inveterate farters: "Fish 'Farts' a Form of Communication?" CBC News, November 6, 2003, https://www.cbc.ca/news/science/fish -farts-a-form-of-communication-1.376991/.

66   described by Aristotle: Aristotle, *Historia Animalium* Book 4, Part 9.

66   were kept largely secret: William J. Broad, "Scientists Fight Navy Plan to Shut Far-Flung Undersea Spy System," *New York Times*, June 12, 1994, https://www.nytimes.com/1994/06/12/us/scientists-fight-navy-plan -to-shut-far-flung-undersea-spy-system.html/.

67   Nobel Prize in 1961: Georg von Békésy, "Concerning the Pleasures of Observing, and the Mechanics of the Inner Ear," Nobel Lecture, December 11, 1961, https://www.nobelprize.org/uploads/2018/06/bekesy -lecture.pdf/.

67   low sounds of 180 Hz: Hanlon and Budelmann, "Why Cephalopods Are Probably Not 'Deaf.'"

67   That would change soon. In 1963: Tavolga, *Marine Bio-Acoustics*, 147–152.

68   local fish hearing tests: William Tavolga and Jerome Wodinsky, "Auditory Capacities in Fishes: Pure Tone Thresholds in Nine Species of Marine Teleosts," *Bulletin of the AMNH* 126, no. 2 (1963): 179–239.

68   fish heard best at around: Tavolga and Wodinsky, "Auditory Capacities."

69   Beefeater gin: Interview with Arthur Popper, 2022.

69   "audio-ichthyotron": W. N. Tavolga, "The Audio-Ichthyotron: The Evolution of an Instrument for Testing the Auditory Capacities of Fishes," *Transactions of the New York Academy of Sciences* 28 (1966): 706–712.

69   Doug Webster: "Douglas Barnes Webster," Legacy.com, accessed September 13, 2023, https://obits.nola.com/us/obituaries/nola/name /douglas-webster-obituary?id=12616345/.

69   "the more interesting way up to his lab": Interview with Arthur Popper, 2022.

70   cave-fish audiograms: Arthur N. Popper, "Auditory Capacities of the Mexican Blind Cave Fish (Astyanax Jordani) and Its Eyed Ancestor (Astyanax Mexicanus)," *Animal Behaviour* 18, no. 3 (1970): 552–562, https://doi.org/10.1016/0003-3472(70)90052-7/.

70   studied fish hearing: Doug Daniels, "Life As a Goldfish," *Connecticut College Magazine* (Summer 2018), https://www.conncoll.edu/news/cc -magazine/past-issues/2018-issues/summer-2018/notebook/09-life-as-a -goldfish.html/.

71   In the lake whitefish: Arthur N. Popper, "Ultrastructure of the Auditory Regions in the Inner Ear of the Lake Whitefish," *Science* 192, no. 4243 (June 4, 1976): 1020–1023, https://doi.org/10.1126/science.1273585.

71   shake the fish: R. R. Fay, "The Goldfish Ear Codes the Axis of Acoustic Particle Motion in Three Dimensions," *Science* 225, no. 4665 (1984): 951– 954; description based on my visit to the original table in Washington.

72   "Why Cephalopods Are Probably Not 'Deaf'": Hanlon and Budelmann, "Why Cephalopods Are Probably Not 'Deaf.'"

72   John Walker and Jerry Whitworth: "The Cold War: History of the SOund SUrveillance System (SOSUS)," DOSITS, accessed September 13, 2023, https://dosits.org/people-and-sound/history-of-underwater-acoustics /the-cold-war-history-of-the-sound-surveillance-system-sosus/.

72   The Cold War ended in 1991: Various dates are given in various places, but the dissolution of the Soviet Union was in 1991.

72   looked (much) closer at the cilia: Interview with Steve Simpson, 2021.

73   basic information: Jack W. Bradbury and Sandra L. Vehrencamp, *Animal Communication* (Sunderland, MA: Sinauer Associates, 2011), 358–361.

CHAPTER 4

74   movie studio-cum-theme-park: "Marineland: Where Movie Stars and Marine Life Met," Governor's House Library, May 25, 2021, https://governorshouselibrary.wordpress.com/2021/05/25/marineland -where-movie-stars-and-marine-life-met/; William N. Tavolga, "Fish Bioacoustics: A Personal History," *Bioacoustics* 12, no. 3 (2002): 101– 104, https://doi.org/10.1080/09524622.2002.9753662/.

74   his wife, Margaret: "William N. Tavolga," Legacy.com, accessed September 11, 2023, https://www.legacy.com/us/obituaries/heraldtribune /name/william-tavolga-obituary?id=12442188/.

74   studying the frillfin goby: Tavolga, "Personal History," 102.

76   Off a coastal causeway: the experiment is described in W. N. Tavolga, "Visual, Chemical and Sound Stimuli as Cues in the Sex Discrimina- tory Behaviour of the Gobiid Fish, Bathygobius Soporator," *Zoologica* 41 (1956): 49–64.

76   in its most basic definition: Jack W. Bradbury and Sandra L. Vehrencamp, *Animal Communication* (Sunderland, MA: Sinauer Associates,
     2011).

76   He wanted to untangle all three: Tavolga, "Visual, Chemical and Sound
     Stimuli."

77   This complex ballet: Arthur N. Popper, "Behavior of Bathygobius Soporator," May 26, 2020, https://www.youtube.com/watch?v=deyVaAm
     ZFg8&t=636s&ab_channel=ArthurN.Popper/.

78   BC Fish Sound Project: "BC Fish Sound Project," JASCO Marine Sciences, accessed September 15, 2023, https://www.jasco.com/bc-fish
     -sound-project. This was the first time I heard that fish make noise, pardon the pun.

78   at the right place and time: William Halliday et al., "The Plainfin Midshipman's Soundscape at Two Sites Around Vancouver Island, British
     Columbia," *Marine Ecology Progress Series* (2018), https://doi.org/10
     .3354/meps12730.

78   hums to find a mate: Halliday et al., "Midshipman's Soundscape."

78   hum at night: Halliday et al., "Midshipman's Soundscape."

79   neat rows of photophores: Cassandra Profita, "Scientists Study 'Singing Fish' for Ways to Improve Human Hearing," OPB, July 8, 2018,
     https://www.opb.org/news/article/hearing-loss-midshipman-fish
     -singing-washington-state/.

79   They spend their year: Interview with Joseph Sisneros, November 2021.

79   They settle . . . in pools of retained water: Interview with Joe Sisneros, November 2021; the nest sites I visited were confirmed as intertidal
     rocky beaches.

79   "They bulk up like a bodybuilder": See sonic muscle index in J. A. Sisneros et al., "Morphometric Changes Associated with the Reproductive
     Cycle and Behaviour of the Intertidal-Nesting, Male Plainfin Midshipman Porichthys Notatus," *Journal of Fish Biology* 74 (2009): 18–36,
     https://doi.org/:10.1111/j.1095-8649.2008.02104.x/.

79   swim bladder extensions: Interview with Joe Sisneros, 2021.

80   on summer nights in the mid-'80s: Mackenzie B. Woods, "Singing Fish
     in a Sea of Noise," *Fisheries* 48 (2023): 185–189, https://doi.org/10.1002
     /fsh.10907/.

80   traveled to California: D. G. Zeddies et al., "Sound Source Localization by the Plainfin Midshipman Fish, Porichthys Notatus," *Journal*

of the Acoustical Society of America 127, no. 5 (May 2010): 3104–3113, https://doi.org/10.1121/1.3365261; PMID: 21117759; Interview with Joe Sisneros, 2021.

81   cutoff frequency: Robert J. Urick, *Principles of Underwater Sound, 3rd Edition* (New York: McGraw-Hill, 1983), 13–14; interview with Jim Miller.

82   She sensitizes to his harmonics: J. A. Sisneros and A. H. Bass, "Seasonal Plasticity of Peripheral Auditory Frequency Sensitivity," *The Journal of Neuroscience: The Official Journal of the Society for Neuroscience* 23, no.3 (2003): 1049–1058, https://doi.org/10.1523/JNEUROSCI.23-03 -01049.2003, especially see Figure 9.

82   a paper by his colleague: Halliday et al., "Midshipman's Soundscape."

84   known as the Fish Listener: "The Fish Listener," *The College Today,* The College of Charleston, November 23, 2020, https://today.cofc.edu /2020/11/23/rodney-rountree-the-fish-listener/.

84   a relative of the midshipman: Strictly speaking, the midshipman is a member of the larger toadfish family, Batrachoididae.

84   Cusk-eel are not a single species: J. G. Nielsen et al., Ophidiiform Fishes of the World (Order Ophidiiformes). An Annotated and Illustrated Catalogue of Pearlfishes, Cusk-Eels, Brotulas and Other Ophidiiform Fishes Known to Date (Food and Agriculture Organization, 1999), 125.

85   they heard eels making sounds: David A. Mann, Jeanette Bowers-Altman, and Rodney A. Rountree, "Sounds Produced by the Striped Cusk-Eel Ophidion Marginatum (Ophidiidae) During Courtship and Spawning," *Copeia* no. 3 (1997): 610–612, https://doi.org/10 .2307/1447568.

85   recordings of cusk-eels: James M. Moulton, "Influencing the Calling of Sea Robins (Prionotus Spp.) with Sound," *Biological Bulletin* 111, no. 3 (1956): 393–98. https://doi.org/10.2307/1539146.

85   Stellwagen Bank: Rodney A. Rountree and Francis Juanes, "First Attempt to Use a Remotely Operated Vehicle to Observe Soniferous Fish Behavior in the Gulf of Maine, Western Atlantic Ocean," *Current Zoology* 56, no. 1 (2010): 90–99.

88   the River Project: "River Project Legacy," Hudson River Park, accessed September 12, 2023, https://hudsonriverpark.org/river-project-legacy/.

88   a slew of biological sounds: Katie A. Anderson, Rodney A. Rountree, and Francis Juanes, "Soniferous Fishes in the Hudson River," *Transactions of the American Fisheries Society* 137 (2008): 616–626.

88   freshwater drum: Rodney Rountree and Francis Juanes, "Potential of Passive Acoustic Recording for Monitoring Invasive Species: Freshwater Drum Invasion of the Hudson River Via the New York Canal System," *Biological Invasions* 19 no. 7 (2017): 2075–2088.

89   among the most valuable fish: Erik Stokstad, "Massive Collapse of Atlantic cod didn't leave evolutionary scars," *Science* (April 7, 2021) https://www.science.org/content/article/massive-collapse-atlantic-cod-didn-t-leave-evolutionary-scars

90   The Scottish government wanted to know: Interview with Tony Hawkins, 2022.

90   head of fisheries: "Leading Scientist Retires," IntraFish, updated 11 July 2012, https://www.intrafish.com/fisheries/leading-scientist-retires/1-1-594713/.

90   graduated from the University of Bristol: Interview with Tony Hawkins, 2022; Anthony Hawkins and Colin Chapman, "Food for Thought: Studying the Behaviour of Fishes in the Sea at Loch Torridon, Scotland," *ICES Journal of Marine Science* 77, no. 7–8 (2020): 2423–2431, https://doi.org/10.1093/icesjms/fsaa118/.

90   wartime sonar operator: Interview with Tony Hawkins, 2022.

90   first to record haddock sounds: A. D. Hawkins and C. J. Chapman, "Underwater Sounds of the Haddock Melanogrammus Aeglefinus," *Journal of the Marine Biological Association of the United Kingdom* 46 (1966): 241–247.

91   Loch Torridon: Hawkins and Chapman, "Underwater Sounds."

91   fish come together in the spring: Interview with Tony Hawkins, 2022; A. D. Hawkins, C. Chapman, and D. J. Symonds, "Spawning of Haddock in Captivity," *Nature* 215 (1967): 923–925.

92   pinpointed these aggregations: Licia Casaretto et al., "Locating Spawning Haddock (Melanogrammus Aeglefinus, Linnaeus, 1758) at Sea by Means Of Sound," *Fisheries Research* 154 (2014): 127–134, https://doi.org/10.1016/j.fishres.2014.02.010.

93   The pistol shrimp: Michel Versluis et al., "How Snapping Shrimp Snap: Through Cavitating Bubbles," *Science* 289 (2000): 2114–2117, https://doi.org/10.1126/science.289.5487.2114/.

93   Polychaete worms: Ryutaro Goto, Isao Hirabayashi, A. Richard Palmer, "Remarkably Loud Snaps During Mouth-Fighting by a Sponge-Dwelling Worm," *Current Biology* 29, no.13 (2019): 617–618, https://doi.org/10.1016/j.cub.2019.05.047.

93   fiddler crabs: "How Do Marine Invertebrates Produce Sound?" DOSITS,

accessed September 11, 2023, https://dosits.org/animals/sound-production
/how-do-marine-invertebrates-produce-sounds/.

93   called autocommunication: Bradbury and Veherenkarp, *Animal Communication*, 362.

CHAPTER 5

94   Spallanzani had a problem: Robert Galambos, "The Avoidance of Obstacles by Flying Bats: Spallanzani's Ideas (1794) and Later Theories," *Isis* 34, no 2. (Autumn, 1942): 132–140.

94   the kind of natural historian: Donald Griffin, *Listening in the Dark: The Acoustic Orientation of Bats and Men* (New Haven, CT: Yale University Press, 1958), 58.

95   "There probably occurs to you": Griffin, *Listening in the Dark*, 60.

95   The bat problem went unsolved: Galambos, "Avoidance," 1942.

95   Sir Hiram Maxim's idea: "Sir Hiram Maxim's Plan To Prevent Sea Collisions," *New York Times*, July 28, 1912, https://timesmachine.nytimes
.com/timesmachine/1912/07/28/100589415.html?pageNumber=45/.

96   finally showed how bats use sound: Whitlow W.L. Au, "History of Dolphin Biosonar Research," *Acoustics Today* 11, no. 4 (Fall 2015), https://acousticstoday.org/history-of-dolphin-biosonar-research-by
-whitlow-w-l-au/.

96   The mythological Echo: Ovid, *Metamorphoses, Book 3*, "She who in others' Words her Silence breaks / Nor speaks herself but when another speaks . . ."

97   the so-called Doppler effect: "Doppler effect," DOSITS, accessed September 15, 2023, https://dosits.org/glossary/doppler-effect/.

97   before it sends out the next click: Whitlow W.L. Au, The Sonar of Dolphins (New York: Springer-Verlag, 1993), see Ch. 4.

97   Griffin speculated: William Schevill and Barbara Lawrence, "Food-Finding by a Captive Porpoise (Turnips Truncatus)," *Museum of Comparative, Zoology* 53 (April 6, 1956).

97   For in 1938: Schevill and Lawrence, "Food-Finding."

97   Ronald V. ("Ronnie") Capo: "Sallie O'Hara: Celebrating 75 years of Marineland," *The St. Augustine Record*, November 17, 2015, https://www
.staugustine.com/story/news/2015/11/17/celebrating-75-years
-marineland/16233643007/.

97   began catching dolphins: William A. Schevill, "Evidence for Echolocation by Cetaceans," *Deep Sea Research* 3, no. 2 (1953): 153–154, https://doi.org/10.1016/0146-6313(56)90096-X/.

98   "This behavior calls to mind": Schevill, "Evidence for Echolocation."

98   two fundamentally different groups: Hal Whitehead and Luke Rendell, *The Cultural Lives of Whales and Dolphins* (Chicago: The University of Chicago Press, 2015), 47.

99   several types of sound: Whitehead and Rendell, *Cultural Lives.*

99   "graded" or combined sounds: E. C. Garland, M. Castellote, and C. L. Berchok, "Beluga Whale (Delphinapterus Leucas) Vocalizations and Call Classification from the Eastern Beaufort Sea Population," *Journal of the Acoustical Society of America* 137, no. 6 (June 2015): 3054–3067, https://doi.org/10.1121/1.4919338/.

99   enticing to the U.S. Navy: Sam Ridgway, *The Dolphin Doctor* (Dublin, New Hampshire: Yankee Books, 1987), 40.

99   got naval funding: In Schevill and Lawrence's "Food Finding" paper, the acknowledgments list support from the Office of Naval Research.

99   Winthrop Kellogg: W. N. Kellogg and Robert Kohler, "Reactions of the Porpoise to Ultrasonic Frequencies," *Science* 116 (1952): 250–252. https://doi.org/10.1126/science.116.3010.250/.

99   By 1956 Schevill and Lawrence watched: Schevill and Lawrence, "Food-Finding."

100   rubber suction cups: Au, "History of Dolphin Biosonar."

100   William McLean: Ridgway, *Dolphin Doctor,* 39; that he read that man would communicate with dolphins, see John C. Lilly, MD., *Man and Dolphin* (New York: Doubleday and Co, 1961).

100   kept bottlenose dolphins for decades: By this point, Marine Studios had been operating for some twenty-five years.

100   lack sweat glands: "The Dolphins That Joined the Navy," Naval History and Heritage, YouTube video uploaded January 13, 2012, https://www.youtube.com/watch?v=vw0wM2KdZE0&ab_channel=Naval HistoryandHeritage/.

100   broken bones: Forrest G. Wood, *Marine Mammals and Man* (New York: Robert B. Luce Inc., 1973), 37.

101   "When released": Wood, *Marine Mammals,* 41–42.

102   Naval Ocean Systems Center field station: Au, "History of Dolphin Biosonar."

102   when echolocation evolved: Nicholas D. Pyenson, "The Ecological Rise of Whales Chronicled by the Fossil Record," *Current Biology 27* (June 5, 2017): 558–564.

102   warmer than today: Fengyuan Li and Shuqiang Li, "Paleocene–Eocene

and Plio–Pleistocene Sea-Level Changes As 'Species Pumps' in Southeast Asia: Evidence from Althepus Spiders," *Molecular Phylogenetics and Evolution* 127 (2018): 545–555, https://doi.org/10.1016/j.ympev.2018.05.014.

102   150 meters higher: Li and Li, "Paleocene–Eocene."

102   the Great Pyramid of Giza: "Great Pyramid of Giza," *Encyclopedia Britannica*, last updated June 16, 2023, https://www.britannica.com/place /Great-Pyramid-of-Giza.

102   All of Florida: Florence Snyder, "Britton Hill: Florida's Highest Natural Point," Visit Florida, accessed September 13, 2023, https://www .visitflorida.com/travel-ideas/articles/arts-history-britton-hill-highest -point-florida/.

102   One early ancestor of modern whales: Pyenson, "Ecological Rise of Whales."

103   length of the cochlea: Geoffrey A. Manley and Jennifer Clack, "An Outline of the Evolution of Vertebrate Hearing Organs," in Manley, Popper, and Fay, *Evolution*, 2004, 14.

103   25 or 30 million years ago: Pyenson, "Ecological Rise of Whales."

103   shrews and mice, and even swiftlet birds: Jo Price, "What Is Echolocation and Which Animals Use It?" *Discover Wildlife Magazine*, May 19, 2022, https://www.discoverwildlife.com/animal-facts/what-is-echolocation/.

103   people who have lost their sight: Daniel Kish, "How I Use Sonar to Navigate the World," TED, filmed March 2015, video, https://www .ted.com/talks/daniel_kish_how_i_use_sonar_to_navigate_the _world?language=en/.

103   Nicholas Pyenson: Interview with Nick Pyenson, 2022.

105   A click is "broadband": Whitlow W.L. Au, *The Sonar of Dolphins* (New York: Springer-Verlag, 1993), 115–120 for an overview of clicks.

105   Cupping the lips: Interview with Nick Pyenson, 2022.

105   "extremely toxic fatty acid chains": Interview with Nick Pyenson, 2022.

106   first things to fall out of the skull: Interview with Nick Pyenson.

106   some of the densest tissue: Interview with Nick Pyenson.

107   The first dolphin audiograms: Au, "History of Dolphin Biosonar."

108   One beluga whale: Whitlow W.L. Au et al., "Demonstration of Adaptation Beluga Whale Signals," *The Journal of the Acoustical Society of America* 77, no. 2 (March 1985): 726–730, https://doi.org/10.1121/1.392341/.

108   Scylla: Wood, *Marine Mammals*, 75.

108   Doris: Wood, *Marine Mammals*, 77–78.

108   ultrasound frequencies vary: Sarah Catchpoole, "Ultrasound Frequencies,"

Radiopaedia, April 20, 2023, https://radiopaedia.org/articles/ultrasound -frequencies/.

109 in deaf humans: Liam J. Norman and Lore Thalor, "Retinotopic-Like Maps of Spatial Sound in Primary 'Visual' Cortex of Blind Human Echolocators," *Proceeding of the Royal Society B: Biological Sciences* 286 (2019), https://doi.org/10.1098/rspb.2019.1910/.

109 ten-meter-high tide: "Tides Today and Tomorrow in Anchorage, AK," US Harbors, accessed September 12, 2023, https://www.usharbors.com /harbor/alaska/anchorage-ak/tides/.

109 fourth-largest: "Frequently Asked Questions," National Oceanic and Atmospheric Administration, accessed September 12, 2023, https:// tidesandcurrents.noaa.gov/faq.html#08. I'm going here by regions, with the three highest being Bay of Fundy, Ungava Bay, Bristol, and Cook Inlet.

110 H. P. Lovecraft's *Dagon*: H. P. Lovecraft, *Necronomicon: The Best Weird Tales of HP Lovecraft* (London: Gollancz, 2008), 5–6.

110 second-busiest cargo airport: "Airport Facts," Alaska Department of Transportation and Public Facilities, Ted Stevens Anchorage International Airport, accessed September 15, 2023, https://dot.alaska.gov/anc /about/facts.shtml/.

111 "sea canaries": Marie P. Fish and William H. Mowbray, "Production of Underwater Sound by the White Whale or Beluga, Delphinapterus Leucas (Pallas)," *Journal of Marine Research* 20, no. 2 (1962), https://elischolar.library.yale.edu/journal_of_marine_research/982/.

111 about 12 feet: "Beluga Whale (Delphinapterus Leucas)," Alaska Department of Fish and Game, accessed September 13, 2023, https://www .adfg.alaska.gov/index.cfm?adfg=beluga.main/.

111 can reach 18 feet: "Introduction," Pacific Mammal Research, accessed September 10, 2023, https://pacmam.org/wp/marine-mammal -highlights-beluga-2/.

111 1,300 belugas: "Recovery Plan for the Cook Inlet Beluga Whale (Delphinapterus Leucas)," United States National Marine Fisheries Service, Alaska Regional Office, Protected Resources Division, December 2016, https://repository.library.noaa.gov/view/noaa/15979/.

112 bespoke hydrophone moorings: M. Castellote et al., "Beluga Whale (*Delphinapterus Leucas*) Acoustic Foraging Behavior and Applications for Long Term Monitoring," *PLoS ONE* 16, no. 11 (2021), https://doi .org/10.1371/journal.pone.0260485/.

112 how the belugas use clicks and buzzes: Castellote, "Acoustic Foraging."

113   whale oil ran the world: D. Graham Burnett, *The Sounding of the Whale,* (Chicago: University of Chicago Press, 2012), 80.

114   2 million sperm whales: Hal Whitehead and M. Shin, "Current Global Population Size, Post-Whaling Trend and Historical Trajectory of Sperm Whales," *Scientific Reports* 12, no. 19468 (2022), https://doi.org /10.1038/s41598-022-24107-7.

114   strikingly odd: Eric Wagner, "The Sperm Whale's Deadly Call," *Smithsonian Magazine,* December 2011, https://www.smithsonianmag.com /science-nature/the-sperm-whales-deadly-call-94653/.

114   Herman Melville's: Herman Melville, *Moby-Dick* (London: Penguin, 2012).

114   Thomas Beale's: Wagner, "Deadly Call."

114   "When the whale descends": Frederick Debell Bennett, *Narrative of a Whaling Voyage round the globe, from the year 1833 to 1836, comprising sketches of polynesia, California, the Indian Archipelago, etc., with an account of the sperm whale fishery, and the natural history of the climes visited* (London: Richard Bentley, New Burlington Street, 1840), https://whalesite .org/anthology/bennettvoyagev2.htm.

115   "impulsive noises from below": L. Worthington and W. Schevill, "Underwater Sounds Heard from Sperm Whales," *Nature* 180, no. 291 (1957), https://doi.org/10.1038/180291a0.

115   details of sperm-whale sounds: R. H. Backus and W. E. Schevill, "Physeter Clicks" in *Whales, Dolphins, and Porpoises,* ed. K. S. Norris (Berkeley: University of California Press, 1966).

115   "Their astute observations": Ridgway, *Dolphin Doctor,* 197.

115   its squid prey is silent: No squid sounds are on the record.

116   100 times that of the atmosphere: NOAA. How does pressure change with ocean depth? National Ocean Service website, https://oceanservice .noaa.gov/facts/pressure.html. Pressure increases with depth at a rate of 1 atmosphere every 10 meters.1 km = 1000m. 1000/10–100.

116   Two organs nestle: Olga Panagiotopoulou, Panagiotis Spyridis, Hyab Mehari Abraha, David R. Carrier, and Todd C. Pataky, "Architecture of the sperm whale forehead facilitates ramming combat." *Peer J,* 4, e1895 (2016). https://doi.org/10.7717/peerj.1895

116   Satellite tags on sperm whales: L. Irvine et al., "Sperm Whale Dive Behavior Characteristics Derived from Intermediate-Duration Archival Tag Data," *Ecology and Evolution* 28, no. 7 (August 2017): 7822–7837, https://doi.org/10.1002/ece3.3322.

116 beaked whales can dive: Peter J. Austler and Les Watling, "Beaked Whale Foraging Areas Inferred by Gouges in the Seafloor," *Marine Mammal Science* 26, no. 1 (January 2010): 226–233, https://doi.org/10.1111/j.1748-7692.2009.00325.x/.

117 On the inlet: This incident is, naturally, a subjective one: correlation is not causation, after all. Of course I cannot ask the belugas if they indeed turned at the moment of the tide, in response to it. I did take a screenshot of my phone's clock at that moment, but there is really no way to confirm.

CHAPTER 6

118 fission-fusion groups: Valeria Vergara and Marie-Ana Mikus, "Contact Call Diversity in Natural Beluga Entrapments in an Arctic Estuary: Preliminary Evidence of Vocal Signatures in Wild Belugas," *Marine Mammal Science* 35, no. 2 (April 2019): 434–465, https://doi.org/10.1111/mms.12538/.

119 echelon formation: Jeanne Picher-Labrie, "Allocare in Belugas: Still a Mysterious Behaviour," Whales Online, a GREMM project, https://baleinesendirect.org/en/allocare-in-belugas-still-a-mysterious-behaviour/.

120 nurses for up to two years: Cory J. D. Matthews and Steven H. Ferguson, "Weaning Age Variation in Beluga Whales (*Delphinapterus leucas*)," *Journal of Mammalogy* 96, no. 2 (April 2015): 425–437, https://doi.org/10.1093/jmammal/gyv046/.

121 fifteen-year-old Aurora: Valeria Vergara, *Acoustic Communication and Vocal Learning in Belugas (Delphinapterus Leucas)*, PhD dissertation (University of British Columbia, 2011), 21, https://open.library.ubc.ca/media/stream/pdf/24/1.0071602/1.

121 made his first sounds: Valeria Vergara and Lance Barrett-Lennard, "Vocal Development in a Beluga Calf (*Delphinapterus leucas*)," *Aquatic Mammals* 34, no. 1 (January 2008): 123–143, https://doi.org/10.1578/AM.34.1.2008.123/.

121 young monkeys: Vergara and Barrett-Lennard, "Vocal Development."

123 Contact calls are not unique: Noriko Kondo and Shigeru Watanabe, "Contact Calls: Information and Social Function," *Japanese Psychological Research* 51, no. 3 (2009): 197–208, https://doi.org/10.1111/j.1468-5884.2009.00399.x 2009/.

123 a "brat": "Young Beluga Whale Dies Unexpectedly at BC Aquarium," *Seattle Times*, https://www.seattletimes.com/seattle-news/young-beluga-whale-dies-unexpectedly-at-bc-aquarium/.

124   shade into each other: E. C. Garland, M. Castellote, and C. L. Berchok, "Beluga Whale (Delphinapterus Leucas) Vocalizations and Call Classification from the Eastern Beaufort Sea Population," *Journal of the Acoustical Society of America* 137, no. 6 (June 2015): 3054–3067, https://doi.org/10.1121/1.4919338.

125   unfortunate event in 1999: Valeria Vergara, Robert Michaud, and Lance Barrett-Lennard, "What Can Captive Whales Tell Us About their Wild Counterparts? Identification, Usage, and Ontogeny of Contact Calls in Belugas (Delphinapterus Leucas)," *International Journal of Comparative Psychology* 23, no. 3. (2010), https://escholarship.org/uc/item/4gt03961.

125   killer whale near Vancouver: Sheena Goodyear, with files from CBC News' Roshini Nair, "This Orca Mother Has Been Holding Her Dead Calf Afloat for Days," CBC News, *As It Happens,* July 27, 2018, https://www.cbc.ca/radio/asithappens/as-it-happens-friday-edition-1 .4731059/this-orca-mother-has-been-holding-her-dead-calf-afloat-for -days-1.4731063/.

125   unique to each animal: Laela S. Sayigh et al., "Bottlenose Dolphin Mothers Modify Signature Whistles in the Presence of Their Own Calves," *Proceedings of the National Academy of Sciences* 120, no. 27 (June 26, 2023), https://doi.org/10.1073/pnas.230026212/.

126   world's longest-running study: Interview with Laela Sayigh, 2022; "The Sarasota Dolphin Research Program: Our Approach to Helping Dolphins," Sarasota Dolphin Research Program, accessed September 11, 2023, https://sarasotadolphin.org/.

126   resident dolphins had unique whistles: Melba C. Caldwell and David K. Caldwell, "The Whistle of the Atlantic Bottlenosed Dolphin (*Tursiops truncatus*)—Ontogeny" in H. E. Winn and B. L. Olla, eds., *Behaviour of Marine Animals* (Boston: Springer, 1979), 369–401.

126   Melba and David Caldwell: Laela S. Sayigh and Vincent M. Janik, "Signature Whistles" in *Encyclopedia of Marine Mammals (Second Edition),* eds. William F. Perrin, Bernd Würsig, and J.G.M. Thewissen (Cambridge, MA: Academic Press, 2009), 1014–1016, https://doi.org/10 .1016/B978-0-12-373553-9.00235-2.

127   skyrockets to about 90 percent: Interview with Laela Sayigh, 2022.

127   seen some patterns: Interview with Laela Sayigh, 2022.

128   dolphins *learn* these whistles: P. L. Tyack and L. S. Sayigh, "Vocal learning in Cetaceans" in *Social Influences on Vocal Development,* eds. C. T.

Snowdon and M. Hausberger (Cambridge, UK: Cambridge University Press, 1997), 208–233; Stephanie L. King and Vincent Janik, "Bottlenose Dolphins Can Use Learned Vocal Labels to Address Each Other," *Proceedings of the National Academy of Sciences* 110, no. 32 (August 2013): 13216–13221, https://doi.org/10.1073/pnas.1304459110; correspondence with Laela Sayligh, 2022. See also V. M. Janik and M. Knörnschild, "Vocal Production Learning in Mammals Revisited," *Philosophical Transactions of the Royal Society of London B: Biological Sciences* 376, no. 1836 (October 20201), https://doi.org/10.1098/rstb.2020.0244.

129 use each other's signature whistles: King and Janik, "Bottlenose Dolphins."

129 *by-product distinctiveness*: Laela Sayigh et al., "Selection Levels on Vocal Individuality: Strategic Use or Byproduct," *Current Opinion in Behavioral Sciences* 46 (2022), https://doi.org/10.1016/j.cobeha.2022.101140.

130 61 percent of the calls: Vergara and Mikus, "Contact Call Diversity."

131 tower on Kamouraska Island: Interview with Jacqueline Aubin and Valeria Vergara, 2022; Valeria Vergara, @Marine_Valeria, "Day 2. Early start to better secure the tower . . . and then magic happened, with belugas visiting the island, the wind cooperating for some drone flights, and great photo ID sessions for @JaclynAubin's exciting PhD study!! #adayontheresearchisland @oceanwise, @GREMM_, @ROMMrdl," Tweet, @Marine_Valeria, July 20, 2021.

131 live in all the oceans of the world: "Killer Whale," NOAA Fisheries, accessed September 10, 2023, https://www.fisheries.noaa.gov/species/killer-whale/.

132 5,000 live in the North Pacific: Interview with Volker Deecke, 2022.

132 Killer whales are matrilineal: Hal Whitehead and Luke Rendell, *The Cultural Lives of Whales and Dolphins* (Chicago: The University of Chicago Press, 2015), 126-127.

133 There are three "ecotypes": NOAA Fisheries, "Killer Whale."

133 offshore ecotype: "Offshore Killer Whales," The Georgia Strait Alliance, accessed September 10, 2023, https://georgiastrait.org/work/species-at-risk/orca-protection/killer-whales-pacific-northwest/offshore-killer-whales/.

133 Transient killer whales: John K. B. Ford, "Vocal Traditions Among Resident Killer Whales (*Orcinus Orca*) in Coastal Waters of British Columbia," *Canadian Journal of Zoology* 69 (1991).

134 Michael Bigg noticed that he could tell: Michaela Ludwig, "BC's Pioneer of Killer Whale Research," *British Columbia Magazine,* November 4, 2016, https://www.bcmag.ca/bcs-pioneer-of-killer-whale-research/.

134 six-year study: JKB Ford, "A Catalogue of Underwater Calls Produced by Killer Whales (Orcinus Orca) in British Columbia," *Canadian Data Report of Fisheries and Aquatic Sciences* (1987): 633.

134 tens of kilometers: Interview with Volker Deecke, 2022.

135 Pikas do this: Preston Somers, "Dialects in Southern Rocky Mountain Pikas, Ochotona Princeps (Lagomorpha)," *Animal Behaviour* 21, no. 1 (1973): 124–137, https://doi.org/10.1016 S0003-3472(73)80050-8.

135 some songbirds: Interview with Volker Deecke, 2022.

135 When he was done: Ford, "A Catalogue"; Ford, "Vocal Traditions."

135 "This difference is readily apparent": Ford, "A Catalogue."

136 entirely separate menus: Volker B. Deecke, Peter J. B. Slater, and John K. B. Ford, "Selective Habituation Shapes Acoustic Predator Recognition in Harbour Seals," *Nature* 420 (November 2002), 171–173.

137 (peak at lower frequencies): The peak of dolphin clicks is around 100–120 Hz, depending on the species, while the peak frequency of a killer whale click is between 12 and 19 kHz: Amanda A. Leu et al., "Echolocation Click Discrimination for Three Killer Whale Ecotypes in the Northeastern Pacific," *Journal of the Acoustical Society of America* 151 (2002): 3197, https://doi.org/10.1121/10.0010450/.

137 fifty different pulsed calls: Interview with Volker Deecke, 2022.

138 pulsed calls seemed to change over time: Volker B. Deecke, John K. B. Ford, Paul F. Spong, "Dialect Change in Resident Killer Whales: Implications for Vocal Learning and Cultural Transmission," *Animal Behaviour* 60 (2000): 629–638, https://doi.org/10.1006/anbe.2000.1454/.

138 live for a long time: NOAA Fisheries, "Species Directory: Killer Whale," accessed September 11, 2023, https://www.fisheries.noaa.gov/species /killer-whale#:~:text=Lifespan%20%26%20Reproduction,10%20 and%2013%20years%20old.

139 in the North Atlantic: Ana Selbmann et al., "A Comparison of Northeast Atlantic Killer Whale (*Orcinus Orca*) Stereotyped Call Repertoires," *Marine Mammal Science* 37, no. 1 (January 2021): 268–289, https://doi .org/10.1111/mms.12750/.

139 low-pitched sound when they herd herring: Whitehead and Rendell, *Cultural Lives,* 140; Malene Simon et al., "Icelandic Killer Whales Orcinus Orca Use a Pulsed Call Suitable for Manipulating the Schooling

Behavior of Herring Clupea Harengus," *Bioacoustics* 16, no. 1 (2006): 57–74, https://doi.org/10.1080/09524622.2006.9753564; Interview with Volker Deecke, 2022.

140 "a system of communication": "Language," New World Encyclopedia, accessed September 12, 2023, "https://www.newworldencyclopedia.org /entry/Language.

140 "Language is a purely human": Edward Sapir, "Introductory: Language Defined," Chapter 1 in *Language: An Introduction to the Study of Speech* (New York: Harcourt, Brace & World, 1921), 3–23, https://brocku.ca /MeadProject/Sapir/Sapir_1921/Sapir_1921_01.html/.

140 "Language is a special form": Forrest G. Wood, *Marine Mammals and Man* (New York: Robert B. Luce Inc., 1973), 101.

140 "The limits of my language": L. C. Sterling, "The Limits of My Language Mean the Limits of My World," *Medium*, November 1, 2014, https://medium.com/@lcsterling/the-limits-of-my-language-mean-the -limits-of-my-world-68b94fc1d119/.

141 language requires *syntax*: Interview with Hal Whitehead, 2022: Whitehead and Rendell, *Cultural Lives*, 289.

141 The bit rate of whale calls: Interview with Volker Deecke, 2022.

CHAPTER 7

142 Humpback whales travel back and forth: Hal Whitehead and Luke Rendell, *The Cultural Lives of Whales and Dolphins* (Chicago: The University of Chicago Press, 2015), 68.

142 fourteen distinct populations: "Humpback Whale," NOAA Fisheries, accessed September 12, 2023, https://www.fisheries.noaa.gov/species /humpback-whale/.

143 Palisades station: "St. David's Headquarters," SOFAR Bermuda, accessed September 12, 2023, https://sofarbda.org/st-davids-headquarters .html/.

143 snaked out and down: David Rothenberg, *Thousand Mile Song: Whale Music in a Sea of Sound* (New York: Basic Books, 2008), 14.

143 Frank Watlington: D. Graham Burnett, *The Sounding of the Whale* (Chicago: The University of Chicago Press, 2012), 635.

143 he didn't tell anyone: "The Discovery," Ocean Alliance, accessed September 10, 2023, https://whale.org/humpback-song/.

143 young acoustician named Scott McVay: Burnett, *Sounding*, 634.

144 university's birdsong lab: Burnett, *Sounding*, 636.

144 long black-and-white printouts: See the *Science* cover, Roger S. Payne and Scott McVay, "Songs of Humpback Whales," *Science* 173 (1971): 585–597, https://doi.org/10.1126/science.173.3997.585/.

144 what counted as "song" in animals: W. B. Broughton, "Methods in Bio-Acoustic Terminology" in *Acoustic Behaviour of Animals*, ed. R. G. Busnel (New York: Elsevier, 1963), 3–24.

144 released an album: Burnett, *Sounding*, 629.

145 St. Lawrence belugas: I came across a copy of this on display at the New Bedford Whaling Museum (which has also published a wonderful volume of Schevill's and William Watkins's work, *The Voices of Marine Mammals*, which also includes the album *Whale and Porpoise Voices*, with recordings from 1962).

145 *Sounds of Sea Animals*: W. N. Kellogg, *Sounds of Sea Animals: Vol. 2 Florida*, Discogs, accessed September 12, 2023, https://www.discogs.com/release/3662216-W-N-Kellogg-Sounds-Of-Sea-Animals-Vol-2-Florida/.

145 published their paper . . . *Science*: Payne and McVay, "Songs."

145 the medieval musical notation of "neumes": Rothenberg, *Thousand Mile Song*, 139–140.

145 gathering momentum: Pinpointing the beginning of something as complex as the conservation movement isn't possible, but Carson's *Silent Spring* in 1962 is credited as a significant point by, among others, organizations such as the National Resources Defense Council, https://www.nrdc.org/stories/story-silent-spring/.

145 proposed a moratorium: Burnett, *Sounding*, 16.

146 "vocal or instrumental sounds": "Song," Oxford English Dictionary, https://www.oed.com/search/dictionary/?scope=Entries&q=song/.

146 less than 1 bit per second: Interview with Volker Deecke, 2022: Also for example, Ryuji Suzuki, John R. Buck, and Peter L. Tyack, "Information Entropy of Humpback Whale Songs," *Journal of the Acoustical Society of America* 119 (2006): 1849–1866.

146 "Like human music": Rothenberg, *Thousand Mile Song*, 132.

146 there is structure there: Payne and McVay, "Songs."

147 "Sequences enable": Nolan Gasser, *Why You Like It* (New York: Flatiron Books, 2022), p. 60.

147 the Paynes traveled the world: Roger Payne, *Among Whales* (New York: Scribner, 1995), 62, and the rest of the chapter "Living Among Whales in Patagonia."

147 over time, these songs *evolved*: Payne, *Among Whales*, 147; K. Payne, P. Tyack, and R. S. Payne, "Progressive Changes in the Songs of Humpback Whales (*Megaptera Novaeangliae*): A Detailed Analysis of Two Seasons in Hawaii" in *Communication and Behavior of Whales*, ed. R. Payne (Boulder, CO: Westview Press, 1983), 9–57.

147 Arranging the themes in a wheel: Rothenberg, *Thousand Mile Song*, 154.

147 Douglas Cato and assistants: Rothenberg, *Thousand Mile Song*, 147.

147 something striking unfold: Michael Noad et al., "Cultural Revolution in Whale Songs," *Nature* 408, no. 537 (2000), https://doi.org/10.1038/35046199/.

148 topped the Billboard 100: *Billboard*, December 27, 1997–January 3, 1998.

148 Puff Daddy and Faith Hill's: *Billboard*, December 27, 1997–January 3, 1998.

148 cast as a tribute: Todd S. Purdum, "Rapper Is Shot to Death in Echo of Killing Six Months Ago," *New York Times*, March 10, 1997, https://www.nytimes.com/1997/03/10/us/rapper-is-shot-to-death-in-echo-of-killing-6-months-ago.html/.

148 "like the typical": Whitehead and Rendell, *Cultural Lives*, 84.

149 songs drifted eastward: Ellen C. Garland et al., "Dynamic Horizontal Cultural Transmission of Humpback Whale Song at the Ocean Basin Scale," *Current Biology* 21, no. 8 (April 26, 2011), 687–691.

149 "the way we do things": Interview with Valeria Vergara, 2022.

149 "information or behavior": Whitehead and Rendell, *Cultural Lives*, 12.

149 *rhymed*: L. N. Guinee and K. B. Payne, "Rhyme-Like Repetitions in Songs of Humpback Whales," *Ethology* 79, no. 4 (1988), 295–306, https://doi.org/10.1111/j.1439-0310.1988.tb00718.x/.

150 The obvious explanation: Interview with David Mellinger, 2021 and 2022.

150 male coordination: James D. Darling, Meagan E. Jones, and Charles P. Nicklin, "Humpback Whale Songs: Do They Organize Males During the Breeding Season?" *Behaviour* 143, no. 9 (2006): 1051–1101, http://www.jstor.org/stable/4536395.

150 a kind of sonar: Eduardo Mercado III, "The Sonar Model for Humpback Whale Song Revised," *Frontiers in Psychology* 9, no. 1156 (2018), https://doi.org/10.3389/fpsyg.2018.01156/.

150 song is supposed to travel a long way: In animal communication theories, a signal may be repeated to ensure that it's received, described for example in Jack W. Bradbury and Sandra L. Vehrencamp, *Animal*

Communication (Sunderland, MA: Sinauer Associates, 2011), 469, in terms of redundancy.

150   a week at the Oregon Zoo: Jane E. Brody, "Scientist at Work: Katy Payne," *New York Times,* November 9, 1993, https://www.nytimes .com/1993/11/09/science/scientist-at-work-katy-payne-picking-up -mammals-deep-notes.html/.

150   easily travels 2 kilometers: K. B. Payne, W. R. Langbauer, and E. M. Thomas, "Infrasonic Calls of the Asian Elephant (*Elephas maximus*)," *Behavioral Ecology and Sociobiology* 18 (1986): 297–301, https://doi.org /10.1007/BF00300007; William R. Langbauer et al., "African Elephants Respond to Distant Playbacks of Low-Frequency Conspecific Calls," *Journal of Experimental Biology* 157, no. 1 (May 1991): 35–46, https://doi.org/10.1242/jeb.157.1.35/.

151   masters of songs: Only baleen whales produce "song," but not all baleen whales do.

151   The blue whale: "Baleen Whales," The Center for Coastal Studies, accessed September 12, 2023, https://coastalstudies.org/connect-learn /stellwagen-bank-national-marine-sanctuary/marine-mammals /cetaceans/baleen-whales/.

151   so huge: D. E. Cade, et al., "Minke Whale Feeding Rate Limitations Suggest Constraints On The Minimum Body Size For Engulfment Filtration Feeding," *Nature Ecology and Evolution* 7 (2023): 535–546, https://doi.org/10.1038/s41559-023-01993-2/.

151   because they are in the water: Whitehead and Rendell, *Cultural Lives,* 53.

152   ocean-pervading 20-Hz "blips": W. E. Schevill, W. A. Watkins, and R. H. Backus, "The 20-Cycle Signals and Balaenoptera (Fin Whales)" in *Marine Bio-Acoustics: Proceedings of a Symposium Held at the Lerner Marine Laboratory, Bimini, Bahamas, April 11–13, 1963,* ed. William N. Tavolga (London: Pergamon Press, 1964), 147–152.

152   Blue whales call very low: "Blue Whale: Sounds of Blue Whale" DOSITS, accessed September 11, 2023, https://dosits.org/galleries /audio-gallery/marine-mammals/baleen-whales/blue-whale/.

152   Sei whales, the mysterious: "Sei Whale," DOSITS, accessed September 11, 2023, https://dosits.org/galleries/audio-gallery/marine-mammals /baleen-whales/sei-whale/.

152   Fin whale calls are perhaps the simplest: "Fin Whale," DOSITS, accessed September 11, 2023, https://dosits.org/galleries/audio-gallery /marine-mammals/baleen-whales/fin-whale/.

153 In the eastern North Pacific: R. P. Dziak, et al., "A Pulsed-Air Model of Blue Whale B Call Vocalizations," *Scientific Reports* 7, no. 9122 (2017), https://doi.org/10.1038/s41598-017-09423-7/.

153 Their song is a sequence of A and B: See Abstract of Ally Rice et al., "Update on Frequency Decline of Northeast Pacific Blue Whale (Balaenoptera Musculus) Calls," *PLOSOne* 17, no. 14 (2022).

153 shift the frequency lower: Mark A. McDonald, John A. Hildebrand, and Sarah Mesnick, "Worldwide Decline in Tonal Frequencies of Blue Whale Songs," *Endangered Species Research* 9 (2009): 13–21.

153 A call is shifting: Rice et al., "Update."

154 fundamental frequency at 52 Hz: William A. Watkins et al., "Twelve Years of Tracking 52-Hz Whale Calls from a Unique Source in the North Pacific," *Deep Sea Research Part I: Oceanographic Research Papers* 51, no.12 (2004): 1889–1901.

155 captured public imagination: See Leslie Jamieson, "52 Blue" in Leslie Jamieson, *Make It Scream, Make It Burn* (New York: Back Bay Books, 2019), 3–27; Joshua Zeman, dir., *The Loneliest Whale: The Search for 52*, produced by Bleecker Street, https://www.youtube.com/watch?v=uDBZ3pTe4Jg/.

156 a couple of times its body length: Interview with Jim Miller, 2021.

156 only the simplest signals: Eduardo Mercado, "Coding and Redundancy: Man-Made and Animal-Evolved Signals: Jack P. Hailman," *Integrative and Comparative Biology* 48, no.6 (December 2008): 875–876, https://doi.org/10.1093/icb/icn095/.

156 multipath: W. Munk, P. Worcester, and C. Wunsch, *Ocean Acoustic Tomography* (Cambridge, UK: Cambridge University Press, 1995).

157 proposed in the late 1970s: Joshua Horowitz, *War of the Whales* (New York: Simon & Schuster, 2014), 158–159.

157 Munk was an oceanographic forecaster: Kat Galbraith, "Walter Munk: The Einstein of the Oceans," *New York Times*, August 24, 2015, https://www.nytimes.com/2015/08/25/science/walter-munk-einstein-of-the-oceans-at-97.html/.

157 receivers on the other side of oceans: Walter H. Munk et al., "The Heard Island Feasibility Test," *Journal of the Acoustical Society of America* 96, no. 4 (October 1994), https://doi.org/0001-4966/94/96(4)/2330/13/.

158 ("a deaf whale is a dead whale"): Richard C. Paddock, "Undersea Noise Test Could Risk Making Whales Deaf," *Los Angeles Times*, March 22, 1994, https://www.latimes.com/archives/la-xpm-1994-03-22-mn-37069-story.html/.

159 "acoustic herd": R. Payne and D. Webb, "Orientation by Means of Long Range Acoustic Signaling in Baleen Whales," *Annals of the New York Academy of Sciences* (December 1971): 110–141, https://doi.org/10.1111/j.1749-6632.1971.tb13093.x; PMID: 5288850.

159 "no ocean big enough": Payne and Webb, "Songs," Figure 1.

159 opposite sides of oceans: David Brand, "Secrets of whales' long-distance songs are being unveiled by U.S. Navy's undersea microphones—but sound pollution threatens," *Cornell Chronicle*, February 19, 2005, https://news.cornell.edu/stories/2005/02/secrets-whales-long-distance-songs-are-unveiled.

160 Just after 1:30 A.M.: This imagined scene is based on a recording posted by ONC and shared with me as well as posted onto the YouTube channel here: https://www.youtube.com/watch?v=yurmOoHvWkU&ab_channel=NeptuneCanada.

161 pair of ragged claws: T. S. Eliot, "The Love Song of J. Alfred Prufrock," *Gleeditions*, April 17, 2011, www.gleeditions.com/alfredprufrock/students/pages.asp?lid=303&pg=7. Originally published in *Poetry: A Magazine of Verse*, June 1915, 130–135.

162 Cascadia Basin node of Ocean Networks: See Ocean Networks Canada's observatory: https://www.oceannetworks.ca/observatories/physical-infrastructure/cabled-networks/. Image available at Christopher Barnes et al., "Understanding Earth: Ocean Processes using Real-time Data from NEPTUNE, Canada's Widely Distributed Sensor Networks, Northeast Pacific," *Geoscience Canada* 38 (2011): 21–30.

162 The Cascadia Basin Node was built: Interview with Dwight Owens of Ocean Networks Canada, 2021.

162 "fin whales of opportunity": Václav M. Kuna and John L. Nábělek, "Seismic Crustal Imaging Using Fin Whale Songs," *Science* 371 (2021): 731–735, https://doi.org/10.1126/science.abf3962/.

CHAPTER 8

166 the word "absurd": Diane Ackerman, *A Natural History of the Senses* (New York: Random House, 1990), 175.

167 the sun is glinting: BabyWildFilms, KOMO 4 News (ABC Seattle), "US Navy Sonar Blasts Orcas," April 30, 2013, YouTube video: The KOMO Seattle news report seen here shows the weather at the scene: https://www.youtube.com/watch?v=tQ0JPLyYoJk&t=53s&ab_channel=BabyWildFilms/.

167 a pod of killer whales: National Marine Fisheries Service, Office of Protected Resources, "Assessment of Acoustic Exposures on Marine Mammals in Conjunction with USS *Shoup* Active Sonar Transmissions in the Eastern Strait of Juan de Fuca and Haro Strait, Washington, 5 May 2003," January 21, 2005.

167 which Michael Bigg dubbed J-2: Elin Kelsey, "What Happens When an Endangered Whale Pod Loses Its Wise Old Grandma?" *Hakai Magazine*, January 25, 2017, https://hakaimagazine.com/news/what-happens -when-endangered-whale-pod-loses-its-wise-old-grandma/.

167 (Ruffles): "JPod," The Whale Trail, accessed September 12, 2023, https://thewhaletrail.org/wt-species/j-pod/.

167 live capture for aquariums: "Southern Resident Killer Whales," U.S. Environmental Protection Agency, updated June 2021, https://www.epa .gov/salish-sea/southern-resident-killer-whales/.

167 Ken Balcomb: Joshua Horowitz, *War of the Whales* (New York: Simon & Schuster, 2014), 309.

167 David Bain: Interview with David Bain, 2021; National Marine Fisheries Service, "Assessment of Acoustic Exposures."

168 Bain also describes: National Marine Fisheries Service, "Assessment of Acoustic Exposures."

168 the USS *Shoup*: U.S. Navy, "Allegations of Marine Mammal Impacts."

168 trying to head closer to the shore: Interview with David Bain, 2021.

168 testing and training ground: See, for example, the Whiskey Golf testing range near Nanaimo: https://www.navy-marine.forces.gc.ca/assets /NAVY_Internet/docs/en/poesb/hso(e)_7356gr_cfmetr-area_wg -warning_(2016-12-21)_en1.pdf: https://nwtteis.com/About-the-Study -Area/.

168 recorded sonar: Personal communication, 2015.

169 naval report: U.S. Navy, Pacific Fleet, "Report on the Results of the Inquiry into Allegations of Marine Mammal Impacts Surrounding the Use of Active Sonar by USS *SHOUP* (DDG 86) in the Haro Strait on or About 5 May 2003" (2004), Figure 1.

169 Sonar, by nature: Angela D'Amico et al., "Beaked Whale Stranding and Naval Exercises," *Aquatic Mammals* 35, no. 4 (2009): 452–472, esp. 455.

169 can be gruesome: E.C.M. Parsons, "Impacts of Navy Sonar on Whales and Dolphins: Now Beyond a Smoking Gun?" *Frontiers of Marine Science* 4 (September 13, 2017), https://doi.org/10.3389/fmars.2017.00295/.

169 accompanied by reports: M. Simmonds and L. Lopez-Jurado, "Whales and the Military," *Nature* 351, no. 448 (1991), https://doi.org/10.1038/351448a0.

169 a natural occurrence: U.S. Navy, "Allegations of Marine Mammal Impacts"; Department of Conservation, Te Papa Atawhai, "Why Do Marine Mammals Strand?" accessed September 12, 2023, https://www.doc.govt.nz/nature/native-animals/marine-mammals/marine-mammal-strandings/why-do-marine-mammals-strand/.

169 In the 1980s: Simmonds and Lopez-Jurado, "Whales and the Military."

170 beaked whales appear to panic: J. E. Stanistreet et al., "Changes in the Acoustic Activity of Beaked Whales and Sperm Whales Recorded During a Naval Training Exercise Off Eastern Canada," *Scientific Reports* 12 (2022): 1973, https://doi.org/10.1038/s41598-022-05930-4/.

170 ascending too fast: Horowitz, *War of the Whales*, 278–279.

170 Naval reports said: U.S. Navy, "Allegations of Marine Mammal Impacts."

171 Balcomb: Horowitz, *War of the Whales*, 309.

171 Bain, and others: National Marine Fisheries Service, "Assessment of Acoustic Exposures."

171 a tricky word: Anthony D. Hawkins and Arthur N. Popper, "A Sound Approach to Assessing the Impact of Underwater Noise on Marine Fishes and Invertebrates," *ICES Journal of Marine Science* 74, no. 3 (March–April 2017): 635–651, https://doi.org/10.1093/icesjms/fsw205/.

171 powerful, close, and inescapable: Anthony D. Hawkins and Arthur N. Popper, "Assessing the Impact of Underwater Sounds on Fishes and Other Marine Life," *Acoustics Today* (Spring 2014), 30–41.

171 *impulsive*: Hawkins and Popper, "Assessing the Impact."

171 "rise time": Hawkins and Popper, "Assessing the Impact."

172 has a reflex: Paul E. Nachtigall and Alexander Y. Supin, "A False Killer Whale Reduces Its Hearing Sensitivity When a Loud Sound Is Preceded by a Warning," *The Journal of Experimental Biology* 216 (2013), 3062–3070, https://doi.org/10.1242/jeb.085068/.

172 more hearing problems: Gordon Hastie et al., "Effects of Impulsive Noise on Marine Mammals: Investigating Range-Dependent Risk," *Ecological Applications* 29 (2019), e01906. 10.1002/eap.1906.

172 "zones of influence": Peter Mcgregor et al., "Anthropogenic Noise and Conservation" in H. Brumm, ed., *Animal Communication and Noise* (2013): 409–444. 10.1007/978-3-642-41494-7_14.

172 little spiral, the cochlea: Darlene Ketten, "Causes of Hearing Loss in Marine Mammals," DOSITS Webinar, October 5, 2021, https://dosits.org/decision-makers/webinar-series/2021-webinar-series/webinar-hearing-loss/.

173 hearing issues: Ketten, "Causes of Hearing Loss."

173 U.S. Marine Mammal Protection Act: U.S. Congress, "Marine Mammal Protection Act of 1972," U.S. Government Publishing Office, December 26, 2022, https://www.govinfo.gov/app/details/COMPS-1679/.

173 an exemption to the Act: U.S. Congress, "Protection Act," 16–18, 24.

174 grouped animals by hearing range: Southall et al., "Marine Mammal Noise Exposure Criteria: Updated Scientific Recommendations for Residual Hearing Effects," *Aquatic Mammals* 45, no. 2 (2019): 125–232, https://doi.org/10.1578/AM.45.2.2019.125/.

174 haven't been able to study or test them: The first detailed tests were only done in the summer of 2023: "First Successful Hearing Tests Conducted with Minke Whales Will Improve Conservation and Protection of Baleen Whales Globally," *National Marine Mammal Foundation,* updated July 7, 2023, https://www.nmmf.org/our-work/biologic-bioacoustic-research/minke-whale-hearing/.

174 create models: Interview with Nick Pyenson, 2021.

174 minkes, one of the smallest baleen whales: Ian N. Durbach et al., "Changes in the Movement and Calling Behavior of Minke Whales (*Balaenoptera acutorostrata*) in Response to Navy Training," *Frontiers of Marine Science* 9 (July 2021): Sec. Marine Ecosystem Ecology, Volume 8, https://doi.org/10.3389/fmars.2021.660122.

174 If blue whales are feeding: Jeremy A. Goldbogen et al., "Blue Whales Respond to Simulated Mid-Frequency Military Sonar," *The Royal Society Biological Sciences* 280 (August 2013), https://doi.org/10.1098/rspb.2013.0657/.

174 experiments in Norway: L. D. Sivle et al., "Changes in Dive Behavior During Naval Sonar Exposure in Killer Whales, Long-Finned Pilot Whales, and Sperm Whales," *Frontiers of Physiology* 3, no. 400 (October 2012), https://doi.org/10.3389/fphys.2012.00400; PMID: 23087648; PMCID: PMC3468818/.

175 tested low-frequency sonar: Arthur N. Popper et al., "The Effects of High-Intensity, Low-Frequency Active Sonar on Rainbow Trout," *Journal of the Acoustical Society of America,* 122 no. 1 (July 2007): 623–635, https://doi.org/10.1121/1.2735115/.

176 subsequently tested sonar on bass, perch, and catfish: Michele B.

Halvorsen et al., "Effects of Low-Frequency Naval Sonar Exposure on Three Species of Fish," *Journal of the Acoustical Society of America* 134 (2013): 205–210, https://doi.org/10.1121/1.4812818/.

176   off the northern coast of Australia: Interview with Robert McCauley, 2021.

176   powerful pulses of sound: Robert C. Gisiner, "Sound and Marine Seismic Surveys," *Acoustics Today* 12, no. 4 (Winter 2016): 10–16, https://acousticstoday.org/wp-content/uploads/2018/08/Sound-and -Marine-Seismic-Surveys-Robert-C.-Gisiner.pdf/.

176   well-known humpback-whale nursery: Lars Bejder et al., "Low Energy Expenditure and Resting Behaviour of Humpback Whale Mother-Calf Pairs Highlights Conservation Importance of Sheltered Breeding Areas," *Nature Scientific Reports* 9, no. 771 (2019), https://doi.org/10 .1038/s41598-018-36870-7/.

176   gallons of yogurt-thick milk: The milk is very high in fat depending on the species, and I have heard comparisons to everything from sour cream to toothpaste.

176   path out of the gulf: R. D. McCauley et al., "The Response of Humpback Whales (*Megaptera novangeliae*) to Offshore Seismic Survey Noise: Preliminary Results of Observations About a Working Seismic Vessel and Experimental Exposures," *The APPEA Journal* 38 (1998), 692–707, https://doi.org/10.1071/AJ97045.

177   "We ended up being out there": Interview with Robert McCauley, 2021.

177   "seismic vessel was doing": McCauley et al., "Response."

177   tiny photosynthetic algae: "Bioluminescence," National Geographic Education, accessed September 12, 2023, https://education.national geographic.org/resource/bioluminescence/ (see "Other Bioluminesence").

177   might disturb large humpbacks: McCauley et al., "Response."

177   Plankton come in all shapes: "Plankton," National Geographic Education, accessed September 12, 2023, https://education.nationalgeographic .org/resource/plankton/.

178   ply the same areas for months: Interview with Robert McCauley, 2021.

178   a small number of industry surveying companies: Interview with Jayson Semmens, 2021.

178   began on land: Susan Schlee, *The Edge of an Unfamiliar World* (New York: E.P. Dutton & Co, 1973), 325–326.

179   mapped some structures: J. B. Hersey and M. Ewing, "Seismic Reflections From Beneath the Ocean Floor," *Eos, Transactions, Ameri-*

can Geophysical Union 30, no. 1 (1949): 5–14, https://doi.org/10.1029 /TR030i001p00005/.

179   during the 1950s: William E. Schevill, Allyn C. Vine, and Charles Innis, "Memorial to John Brackett Hersey, 1913–1992," *Geological Society of America* (1993): 207–209, https://rock.geosociety.org/net/documents /gsa/memorials/v24/Hersey-JB.pdf/.

179   a more controlled sound source: J. B. Hersey, "Continuous Seismic Reflection Profiling" in *The Sea*, vol. 3, ed. M. N. Hill (New York: Interscience-Wiley, 1963), 47–72.

180   30 percent of global crude oil: The United States Energy Information Association, "Offshore production nearly 30% of global crude oil output in 2015," *Today in Energy*, last updated October 25, 2016, https://www .eia.gov/todayinenergy/detail.php?id=28492#/.

180   Ships dragged long "streamers": "Marine Seismic Surveys: The Search for Oil and Natural Gas Offshore," Canada's Oil and Natural Gas Producers (CAPP), November 2015, https://www.capp.ca/wp-content /uploads/2019/11/Marine_Seismic_Surveys_The_Search_for_Oil _and_Natural_Gas_Offshor-291866.pdf/.

180   four hours just to turn: Interview with Robert McCauley, 2021.

180   The average survey: CAPP, "Marine Seismic Surveys."

180   last-minute chance: Interview with Robert McCauley, 2021.

180   March of 2015: Interview with Robert McCauley, 2021, description of the experiment, also in ref below.

181   knocked out plankton: R. D. McCauley et al., "Widely Used Marine Seismic Survey Air Gun Operations Negatively Impact Zooplankton," *Nature Ecology and Evolution* 1 (2017), https://doi.org/10.1038/s41559-017-0195.

182   oily substances: Interview with Robert McCauley, 2021.

182   Others tried to replicate: David M. Fields et al., "Airgun Blasts Used in Marine Seismic Surveys Have Limited Effects on Mortality, and No Sublethal Effects on Behaviour Or Gene Expression, in the Copepod *Calanus finmarchicus*," *ICES Journal of Marine Science* 76, no.7 (December 2019): 2033–2044, https://doi.org/10.1093/icesjms/fsz126/.

183   More than twenty oil and gas platforms: Interview with Jayson Semmens, 2021.

183   blamed a recent seismic survey: "Scallop Deaths Spark $70m Compo Claim," ABC News, last updated Thursday May 12, 2011, https://www .abc.net.au/news/2011-05-13/scallop-deaths-spark-70m-compo-claim /2710510/.

183 a thorough exam: Ryan D. Day et al., "Assessing the Impact of Marine Seismic Surveys on Southeast Australian Scallop and Lobster Fisheries," Fisheries Research and Development Corporation, University of Tasmania, Hobart, FRDC 2012/008 (2016), http://www.frdc.com.au/ArchivedReports/FRDC%20Projects/2012-008-DLD.pdf/.

184 Semmens says he could *see* the air gun: Interview with Jayson Semmens, 2021.

184 got a lot of "grief": Interview with Jayson Semmens, 2021; R. D. Day et al., "Examining the Potential Impacts of Seismic Surveys on Octopus and Larval Stages of Southern Rock Lobster—PART A: Southern Rock Lobster," FRDC project 2019-051, The Institute for Marine and Antarctic Studies, University of Tasmania, Hobart, Tasmania, 2021.

184 commercial array: Day et al., "Southern Rock Lobster."

184 (Adults live fifteen or even twenty years): Day et al., "Southern Rock Lobster," 26.

185 "The Big K": Maya Wei-Haas, "The Kilogram Is Forever Changed. Here's Why That Matters," *National Geographic,* May 20, 2019, https://www.nationalgeographic.com/science/article/kilogram-forever-changed-why-mass-matters/.

185 decibels are not quite so simple: "Introduction to Decibels," DOSITS, accessed September 10, 2023, https://dosits.org/science/advanced-topics/introduction-to-decibels; Interview with Jim Miller, 2022.

186 20 decibels lower: Gisiner, "Sound and Marine Seismic Surveys."

186 wide range of frequencies: Gisiner, "Sound and Marine Seismic Surveys," Figure 7.

186 four days of testing in the Bass Strait: Interview with Jayson Semmens, 2021; Ryan D. Day et al., "The Impact of Seismic Survey Exposure on the Righting Reflex and Moult Cycle of Southern Rock Lobster (*Jasus Edwardsii*) Puerulus Larvae and Juveniles," *Environmental Pollution* 309 (2022), https://doi.org/10.1016/j.envpol.2022.119699.

187 technological advances: Gisiner, "Sound and Marine Seismic Surveys"; interview with Jayson Semmens, 2021.

187 Many countries: "Moratorium Extended on Oil and Gas Activities in Georges Bank," *Natural Resources Canada,* April 27, 2022, https://www.canada.ca/en/natural-resources-canada/news/2022/04/moratorium-extended-on-oil-and-gas-activities-in-georges-bank.html/; Natalie Pressman, "Feds Extend Restrictions on Arctic Offshore Drilling," CBC News, January 2, 2023, https://www.cbc.ca/news/canada/north/arctic

-offshore-drilling-restrictions-extending-1.6699833; "US Federal Oil and Gas Leasing Hits Historically Low Levels," *Offshore Magazine*, September 11, 2022, https://www.offshore-mag.com/regional-reports /us-gulf-of-mexico/article/14282603/us-federal-oil-and-gas-leasing -hits-historically-low-levels; Morten Butler, "Greenland Bans All Future Oil Exploration, Citing Climate Concerns," *Time*, last updated July 16, 2021, https://time.com/6080933/greenland-bans-oil-exploration/.

188 globally, it's happening less: David H. Johnston, "Four-Dimensional Seismic In The Downturn," Hart Energy, March 27, 2017, https:// www.hartenergy.com/exclusives/four-dimensional-seismic-downturn -29684; and colloquially, the consensus at the Effects of Noise on Aquatic Life 2022 conference was that seismic surveying for oil and gas was generally decreasing.

188 pivoting to another impulsive sound: Effects of Noise on Aquatic Life 2022 conference program saw a sharp uptick in wind farm research.

188 up to 90 meters: Liz Hartman, "Wind Turbines: The Bigger the Better," Office of Energy Efficiency and Renewable Energy, August 24, 2023, https://www.energy.gov/eere/articles/wind-turbines-bigger-better/.

188 easily top 100 meters: Vineyard Wind, currently under construction off Martha's Vineyard, for example, will have hub height of over 100 meters. Blades sweep even higher. "Vineyard Wind: Draft Construction and Operations Plan, Volume 1," https://www.boem.gov/sites/default /files/documents/renewable-energy/Vineyard%20Wind%20COP%20 Volume%20I_Section%203.pdf/.

188 more than 90 percent: "Wind," The International Energy Association, accessed September 12, 2023, https://www.iea.org/energy-system /renewables/wind: "In 2022, of the total 900 GW of wind capacity installed, 93% was in onshore systems, with the remaining 7% in offshore wind farms."

188 Vindeby, in Denmark: "Making Green Energy Affordable," Orsted, accessed September 12, 2023, https://orsted.com/-/media/WWW/Docs /Corp/COM/explore/Making-green-energy-affordable-June-2019.pdf/.

188 injected $3 billion: The White House, "FACT SHEET: Biden Administration Jumpstarts Offshore Wind Energy Projects to Create Jobs," March 29, 2021, https://www.whitehouse.gov/briefing-room/statements -releases/2021/03/29/fact-sheet-biden-administration-jumpstarts -offshore-wind-energy-projects-to-create-jobs/.

189 Block Island Wind Farm: "Block Island Wind Farm," Orsted, accessed

September 10, 2023, https://us.orsted.com/renewable-energy-solutions/offshore-wind/block-island-wind-farm.

189   off the coast of Virginia: "Coastal Virginia Offshore Wind," Dominion Energy, accessed September 10, 2023, https://www.dominionenergy.com/projects-and-facilities/wind-power-facilities-and-projects/coastal-virginia-offshore-wind.

189   (BOEM): J. Amaral et al., "The Underwater Sound from Offshore Wind Farms," *Acoustics Today* 16, no 2 (2020): 13–21, https://doi.org/10.1121/AT.2020.16.2.13.

189   constant low thrum: Dave Seglins and John Nicol, "Wind Farm Health Risks Claimed in $1.5M Suit," CBC News, last updated September 21, 2011, https://www.cbc.ca/news/canada/wind-farm-health-risks-claimed-in-1-5m-suit-1.1044943/.

189   operational noise: F. Thomsen et al., "Effects of Offshore Wind Farm Noise on Marine Mammals and Fish," biola, Hamburg, Germany, on behalf of COWRIE Ltd, 2006, https://tethys.pnnl.gov/sites/default/files/publications/Effects_of_offshore_wind_farm_noise_on_marine-mammals_and_fish-1-.pdf/.

189   some creatures are even attracted: A. Cresci et al., "Atlantic Cod (*Gadus Morhua*) Larvae Are Attracted by Low-Frequency Noise Simulating That of Operating Offshore Wind Farms," *Community Biology* 6, no. 353 (2023), https://doi.org/10.1038/s42003-023-04728-y/.

189   (Shipping firms spend millions): Department of Homeland Security, Acquisition Directorate Research & Development Center, "Vessel Biofouling Prevention and Management Options Report," (2015), https://apps.dtic.mil/sti/tr/pdf/ADA626612.pdf.

189   massive floats: C. M. Wang et al., "Research on Floating Wind Turbines: A Literature Survey, *The IES Journal Part A: Civil & Structural Engineering* 3 no. 4 (2010): 267–277, https://doi.org/10.1080/19373260.2010.517395/.

190   A driven pile: "Pile Driving," DOSITS, accessed September 13, 2023, https://dosits.org/animals/effects-of-sound/anthropogenic-sources/pile-driving/.

190   September of 2015: Bureau of Ocean Energy Management, Office of Renewable Energy Programs, "Field Observations During Wind Turbine Installation at the Block Island Wind Farm, Rhode Island," Final Report to the U.S. Department of the Interior, OCS Study 2019.

190   But where the water is deep: U.S. Dept. of the Interior, Bureau of Ocean

Energy Management, "Comparison of Environmental Effects from Different Offshore Wind Turbine Foundations," OCS Study BOEM 2020-041, Table ES-1.

190 13-degree angle: Amaral et al., "Underwater Sound," Figure 4.

190 between five hundred to more than five thousand strikes: Amaral et al., "Underwater Sound," 15.

190 75 meters: Amaral et al., "Underwater Sound," 14.

190 an acoustic monitoring program: BOEM, "Wind Turbine Installation at the Block Island Wind Farm."

191 maybe some plankton are knocked out: As per the impacts of seismic air guns, a sharp, impulsive noise right at the pile could damage plankton.

191 a salmon or a cod: A. N. Popper and A. D. Hawkins, "An Overview of Fish Bioacoustics and the Impacts of Anthropogenic Sounds on Fishes," *Journal of Fish Biology* 94 (2009): 692–713, Table 2, https://doi.org/10.1111/jfb.13948/.

192 longfin squid: Ian T. Jones, Jenni A. Stanley, and T. Aran Mooney, "Impulsive Pile Driving Noise Elicits Alarm Responses in Squid (*Doryteuthis pealeii*)," *Marine Pollution Bulletin* 150 (2020), https://doi.org/10.1016/j.marpolbul.2019.110792.

192 fin whale: BOEM, "Wind Turbine Installation at the Block Island Wind Farm."

192 seismic waves: "Discern Between Body and Surface Waves, Primary and Secondary Waves, and Love and Rayleigh Waves," *Encyclopedia Britannica*, accessed September 16, 2023, https://www.britannica.com/video/181934/rock-vibrations-Earth-earthquake-waves-P-surface/.

192 evanescent sound waves: Richard Hazelwood and Patrick C. Macey, "Modeling Water Motion near Seismic Waves Propagating Across a Graded Seabed, as Generated by Man-Made Impacts," *Journal of Marine Science and Engineering* 4, no. 3 (2016): 47, https://doi.org/10.3390/jmse4030047.

192 sound using pressure measured above: Arthur N. Popper et al., "Offshore Wind Energy Development: Research Priorities for Sound and Vibration Effects on Fishes and Aquatic Invertebrates," *Journal of the Acoustical Society of America* 151, no. 205 (2022), https://doi.org/10.1121/10.0009237.

192 The sound and its substrate: Y. Jézéquel et al., "Pile Driving Repeatedly Impacts the Giant Scallop (*Placopecten magellanicus*)," *Scientific Reports* 12, no. 15380 (2022), https://doi.org/10.1038/s41598-022-19838-6/.

192   hermit crab: Louise Roberts et al., "Exposure of Benthic Invertebrates to Sediment Vibration: From Laboratory Experiments to Outdoor Simulated Pile-Driving," *Proceedings of Meetings on Acoustics.* 27, no. 1 (2016), https://doi.org/10.1121/2.0000324/.

192   plaice: As Jayson Semmens describes, the seabed visibly fluttering would certainly be tangible to a flatfish. Interview with Jayson Semmens, 2021.

192   The life cycle of entire animal phyla: Adrienne Mason, "The Micro Monsters Beneath Your Beach Blanket," *Hakai Magazine,* March 21, 2016, https://hakaimagazine.com/videos-visuals/micro-monsters-beneath-your-beach-blanket/.

193   may tap into substrate vibrations: Roberts et al., "Exposure of Benthic Invertebrates."

193   one of the oldest communication systems: Peggy S. M. Hill, *Vibrational Communication in Animals* (Cambridge, MA: Harvard University Press, 2008).

CHAPTER 9

194   thirteen species: "The 13 Species," Whales Online: A GREMM Project, accessed September 14, 2023, https://baleinesendirect.org/en/discover/the-species-of-the-st-lawrence/the-13-species/.

195   $15 each: Fisheries and Oceans Canada (DFO), "Recovery Strategy for the Beluga (*Delphinapterus leucas*) St. Lawrence Estuary Population in Canada [Proposed]," Species at Risk Act Recovery Strategy Series, 2011, https://www.canada.ca/en/environment-climate-change/services/species-risk-public-registry/recovery-strategies/beluga-delphinapterus-leucas-st-lawrence-estuary-proposed-2011.html.

196   80 or 90 percent of global trade: "Ocean Shipping and Shipbuilding," OECD, accessed September 2023, https://www.oecd.org/ocean/topics/ocean-shipping/.

196   world's merchant shipping fleet: "Merchant Fleet by Flag of Registration and by Type of Ship, Annual," UNCTAD STAT, accessed September 2023, https://unctadstat.unctad.org/wds/TableViewer/tableView.aspx?ReportId=93/.

196   only about 10 percent of commercial ships: "Number of Ships in the World Merchant Fleet as of January 1, 2022, by Type," *Statista,* accessed September 12, 2023, https://www.statista.com/statistics/264024/number-of-merchant-ships-worldwide-by-type/.

196   half of the world's cargo: Stephanie Nikolopoulos, "Container Shipping:

By the Numbers," Thomas Insights, last updated January 26, 2022, https://www.thomasnet.com/insights/container-shipping-by-the-numbers/.

197   SM *Busan*: "SM Busan," Marine Traffic, accessed September 10, 2023, https://www.marinetraffic.com/en/ais/details/ships/shipid:462822/mmsi:440141000/imo:9312767/vessel:SM_BUSAN/.

197   more than 20,000 TEU capacity: Zahra Ahmed, "Top 20 World's Largest Container Ships in 2023," *Marine Insight*, last updated April 11, 2023, https://www.marineinsight.com/know-more/top-10-worlds-largest-container-ships-in-2019/.

197   traffic increased fourfold: J. Tournadre, "Anthropogenic Pressure on the Open Ocean: The Growth of Ship Traffic Revealed by Altimeter Data Analysis," *Geophysical Research Letters* 41, no. 22 (2014): 7924–7932, https://doi.org/10.1002/2014GL061786/.

197   sound levels have doubled: Mark A. McDonald, John A. Hildebrand, and Sean M. Wiggins, "Increases in Deep Ocean Ambient Noise in the Northeast Pacific West of San Nicolas Island California," *Journal of the Acoustical Society of America* 230, no. 120(2) (August 2006): 711–718, https://doi.org/10.1121/1.2216565/.

197   primarily from the propeller: John A. Hildebrand, "Anthropogenic and Natural Sources of Ambient Noise in the Ocean," *Marine Ecology Progress Series* 395 (2009): 5–20, https://doi.org/10.3354/meps08353/.

198   9 meters: Smita, "8 Biggest Ship Propellers in the World," Marine Insight, March 30, 2019, https://www.marineinsight.com/tech/8-biggest-ship-propellers-in-the-world.

198   broadband: McDonald, Hildebrand, and Wiggins, "Increases in Deep Ocean Ambient Noise."

198   outer rings: Anthony D. Hawkins and Arthur N. Popper, "A Sound Approach to Assessing the Impact of Underwater Noise on Marine Fishes and Invertebrates," *ICES Journal of Marine Science* 74, no. 3 (March–April 2017): 635–651.

199   already been sheared off: Ryan D. Day, et al., "Lobsters with Pre-Existing Damage to Their Mechanosensory Statocyst Organs Do Not Incur Further Damage from Exposure to Seismic Air Gun Signals," *Environmental Pollution* 267 (2020), https://doi.org/10.1016/j.envpol.2020.115478.

199   sensitivity to certain sounds: Talk by David Hannay at Acoustical Society of America Conference in Seattle, 2021, citing Jesse R. Barber,

Kevin R. Crooks, and Kurt M. Fristrup, "The Costs of Chronic Noise Exposure for Terrestrial Organisms," *Trends in Ecology & Evolution* 25, no. 3 (2010): 180–189, https://doi.org/10.1016/j.tree.2009.08.002.

199 cleaner wrasse: Sophie L. Nedelec et al., "Motorboat Noise Disrupts Co-Operative Interspecific Interactions," *Scientific Reports* (2017), https://doi.org/10.1038/s41598-017-06515-2/.

200 twisting, three-dimensional line: Marla Holt et al., "Vessels and Their Sounds Reduce Prey Capture Effort by Endangered Killer Whales (*Orcinus orca*)," *Marine Environmental Research* 170 (2021), https://doi.org/10.1016/j.marenvres.2021.105429.

201 in the waters off Denmark: D. M. Wisniewska et al., "High Rates of Vessel Noise Disrupt Foraging in Wild Harbour Porpoises (*Phocoena phocoena*)," Proceedings of the Royal Society B: Biological Sciences 285 (2018), http://dx.doi.org/10.1098/rspb.2017.2314/.

202 isolated and dwindling population: M. Amundin et al., "Estimating the Abundance of the Critically Endangered Baltic Proper Harbour Porpoise (*Phocoena phocoena*) Population Using Passive Acoustic Monitoring," *Ecology and Evolution* 12 (2022), https://doi.org/10.1002/ece3.8554/.

202 watching a calving ground: V. Vergara et al., "Can You Hear Me? Impacts of Underwater Noise on Communication Space of Adult, Sub-Adult and Calf Contact Calls of Endangered St. Lawrence Belugas (*Delphinapterus leucas*)," *Polar Research* 40 (2021), https://doi.org/10.33265/polar.v40.5521; interview with Valeria Vergara, 2022 and 2023.

203 plainfin midshipmen: Nicholas A. W. Brown et al., "Low-Amplitude Noise Elicits the Lombard Effect in Plainfin Midshipman Mating Vocalizations in the Wild," *Animal Behaviour* 181 (2021): 29–39, https://doi.org/10.1016/j.anbehav.2021.08.025.

203 seven hundred vessels: Interview with Lee Kindberg, 2022.

203 world's largest container-ship company: "MSC Recognized as World's Largest Container Line Surpassing Maersk," *The Maritime Executive*, last updated January 5, 2022, https://maritime-executive.com/article/msc-recognized-as-world-s-largest-container-line-surpassing-maersk/.

203 In the mid-2000s: Interview with Lee Kindberg, 2022.

204 "radical retrofit": V. M. ZoBell et al., "Retrofit-Induced Changes in the Radiated Noise and Monopole Source Levels of Container Ships," *PLoS ONE* 18, no. 3 (2023), https://doi.org/10.1371/journal.pone.0282677/.; interview with Lee Kindberg, 2022.

204 the ECHO program: "Enhancing Cetacean Habitat and Observation (ECHO) Program, Port of Vancouver," accessed September 10, 2023, https://www.portvancouver.com/environmental-protection-at-the -port-of-vancouver/maintaining-healthy-ecosystems-throughout-our -jurisdiction/echo-program/.

205 sound and noise is lost energy: Interview with Lee Kindberg, 2022.

205 boss cap fin: ZoBell, et al., "Retrofit-Induced Changes," 2023; Interview with Lee Kindberg, 2022; Hae-ji, Ju and Jung-sik Choi, "Experimental Study of Cavitation Damage to Marine Propellers Based on the Rotational Speed in the Coastal Waters," *Machines* 10, no. 9 (2022), https://doi.org/10.3390/machines10090793/.

205 monitors a suite of underwater hydrophones: ZoBell et al., "Retrofit-Induced Changes."

206 biggest container ports on the continent: "Overview of California Ports," Legislative Analyst's Office, accessed September 12, 2023, https://lao.ca .gov/Publications/Report/4618/.

206 preliminary data: M. Gassmann, S. M. Wiggins, and J. A. Hildebrand, "Deep-Water Measurements of Container Ship Radiated Noise Signatures and Directionality," *Journal of the Acoustical Society of America* 142, no. 3 (2017): 1563–1574, pmid:28964105.

206 *louder* post-retrofit: ZoBell et al., "Retrofit-Induced Changes."

207 "Lloyd's Mirror" effect: Interview with Vanessa ZoBell, 2022.

207 International Organization for Standardization: Underwater Acoustics: Quantities and procedures for description and measurement of underwater sound from ships, Part 1: Requirements for precision measurements in deep water used for comparison purposes, ISO17208-1:2016 (International Organization for Standardization, 2021), http://www.iso .org/standard/62408.html.

208 require a professional pilot: The minimum requirements are here: "How to Become a Marine Pilot," British Columbia Coast Pilot, https://www .bccoastpilots.com/become-a-marine-pilot/.

208 slowing them down: Krista Trounce et al., "The Effects of Vessel Slowdowns on Foraging Habitat of the Southern Resident Killer Whales," *Proceeding of Meetings on Acoustics* 37, no. 1 (July 2019), https://doi.org /10.1121/2.0001230/.

209 Kindberg likes: Interview with Lee Kindberg, 2022.

210 mandatory slowdown: Transport Canada, "Protecting North Atlantic Right Whales from Collisions with Vessels in the Gulf of St.

Lawrence," last modified September 9, 2023, https://tc.canada.ca/en
/marine-transportation/navigation-marine-conditions/protecting-north
-atlantic-right-whales-collisions-vessels-gulf-st-lawrence/. (Note that
vessels didn't really comply with this very well.)

210   decided to shift the shipping lane: Jakob Tougaard et al., "Effects of Re-
routing Shipping Lanes in Kattegat on the Underwater Soundscape: Report to the Danish Environmental Protection Agency on EMFF project
TANGO," *Scientific Report from DCE–Danish Centre for Environment
and Energy* 63, no. 535 (2023), http://dce2.au.dk/pub/SR535.pdf/.

211   regulates how close: "Marine Activities in the Saguenay–St. Lawrence
Marine Park Regulations," https://parcmarin.qc.ca/wp-content/uploads
/2016/10/ParcMarin-Regulations_v2_www-1.pdf/.

211   "Arctic voyagers": William E. Schevill and Barbara Lawrence, "A Phonograph Record of the Underwater Calls of *Delphinapterus leucas*," Woods
Hole Oceanographic Institution, Woods Hole, Massachusetts, January
1950.

211   The perspective was echoed: Roger S. Payne and Scott McVay, "Songs
of Humpback Whales," *Science* 173 (1971): 585–597, https://doi.org/10
.1126/science.173.3997.585/.

CHAPTER 10

212   bearded seals are an exclusively Arctic species: Department of Fisheries
and Oceans, "Bearded Seal," August 13, 2019, https://www.dfo-mpo.gc
.ca/species-especes/profiles-profils/beardedseal-phoquebarbu-eng.html.

213   1.5 million square kilometers: "Arctic Archipelago," *Encyclopedia
Britannica*, March 10, 2009, https://www.britannica.com/place/Arctic
-Archipelago/.

213   Halliday thought seals were under-researched: Interview with William
Halliday, 2022.

213   baleen bowheads: Their age has been put at 100 years, at least, because
of the recovery of stone whaling harpoon tips in older animals, but they
likely crack 200 years. "Bowhead Whale," NOAA Fisheries, accessed
September 12, 2023, https://www.fisheries.noaa.gov/species/bowhead
-whale/.

213   an elaborate song: Perhaps a subjective judgment, but having listened to
them, I'm apt to agree. K. M. Stafford et al., "Extreme Diversity in the
Songs of Spitsbergen's Bowhead Whales," *Biology Letters* 14, no. 4 (April
2018), https://doi.org/10.1098/rsbl.2018.0056/.

213 Several seals: Protection of the Arctic Marine Environment (PAME), "Underwater Noise in the Arctic: A State of Knowledge Report," Arctic Council, May 2019, 13.

213 "Seal camp": Interview with William Halliday, 2022.

213 The monetary cost: Flying in the Arctic often involves charter flights, the cost of which can easily top CAN $1,000 or even CAN $2,000.

213 designed ship-playback experiments: Interview with William Halliday, 2021.

214 banned non-essential travel: Transport Canada, "Ban on Cruise Ships and Pleasure Craft Due to Covid-19," February 2021, https://tc.canada.ca/en/binder/ban-cruise-ships-pleasure-craft-due-covid-19/.

215 Arctic soundscape is completely different: PAME, "Underwater Noise," 16.

215 the noise of an icebreaker: M. J. Martin et al., "Exposure and Behavioral Responses of Tagged Beluga Whales (*Delphinapterus leucas*) to Ships in the Pacific Arctic," *Marine Mammal Science* 39 (2023): 387–421, https://doi.org/10.1111/mms.12978/.

215 belugas near the Mackenzie: W. D. Halliday et al., "Beluga Vocalizations Decrease in Response to Vessel Traffic in the Mackenzie River Estuary," *Arctic* 72, no. 4 (2019): 337–346, https://www.jstor.org/stable/26867457.

216 recording ambient sound: Interview with William Halliday, 2022.

216 year-round ambient sound: W. D. Halliday et al., "Seasonal Patterns in Acoustic Detections of Marine Mammals Near Sachs Harbour, Northwest Territories," *Arctic Science* 4 (2018): 259–278, https://doi.org/10.1139/AS-2017-0021/.

217 hears fish too: M. K. Pine et al., "Fish Sounds near Sachs Harbour and Ulukhaktok in Canada's Western Arctic," *Polar Biology* 43 (2020): 1207–1216, https://doi.org/10.1007/s00300-020-02701-7/.

217 the cod *Boreagadus saida*: Magnus Aune et al., "Distribution and Ecology Of Polar Cod (*Boreogadus saida*) in the Eastern Barents Sea: A Review of Historical Literature," *Marine Environmental Research* 166 (2021), https://doi.org/10.1016/j.marenvres.2021.105262; https://www.fishbase.se/summary/boreogadus-saida.

217 did not appear to make sounds: Interview with Amalis Riera, 2021.

218 preliminary grunts: A. Riera et al., "Sounds of Arctic Cod (*Boreogadus saida*) in Captivity: A Preliminary Description," *Journal of the Acoustical Society of America* 143, no. 5 (May 2018), https://doi.org/10.1121/1.5035162/; PMID: 29857742.

218   expected to shift profoundly: PAME, "Underwater Noise," 13.
218   four times faster: M. Rantanen, et al., "The Arctic Has Warmed
      Nearly Four Times Faster Than the Globe Since 1979," *Communica-*
      tions Earth and Environment 3, no. 168 (2022), https://doi.org/10.1038
      /s43247-022-00498-3/.
218   ice-free Arctic summer: IPCC, "Regional Fact Sheet—Polar Re-
      gions," Sixth Assessment Report Working Group I—The Physical Sci-
      ence Basis, https://www.ipcc.ch/report/ar6/wg1/downloads/factsheets
      /IPCC_AR6_WGI_Regional_Fact_Sheet_Polar_regions.pdf/.
218   storms already come more frequently: Nikk Ogasa, "Cyclones in the
      Arctic Are Becoming More Intense and Frequent," *Science News,* January
      17, 2023, https://www.sciencenews.org/article/cyclones-arctic-intense
      -frequent-climate/; interview with William Halliday, 2022.
219   arriving sooner: Chelsea Harvey, "As Arctic Sea Ice Melts, Killer Whales
      Are Moving In," *Scientific American,* December 3, 2021, https://www
      .scientificamerican.com/article/as-arctic-sea-ice-melts-killer-whales
      -are-moving-in/.
219   Atlantic cod are moving northward: P. E. Renaud, et al., "Is the pole-
      ward expansion by Atlantic cod and haddock threatening native polar
      cod, *Boreogadus saida*?" *Polar Biology* 35 (2012): 401–412, https://doi.org
      /10.1007/s00300-011-1085-z/.
219   the relative proportion of different copepod species: Niall McGinty et
      al., "Anthropogenic Climate Change Impacts on Copepod Trait Bio-
      geography," *Global Change Biology* 27 (2021): 1431–1442, https://doi.org
      /10.1111/gcb.15499/.
219   Humans have lived in the Arctic: "Inuit Nunangat," Indigenous Peo-
      ples' Atlas of Canada, accessed September 12, 2023, https://indigenous
      peoplesatlasofcanada.ca/article/inuit-nunangat/.
219   70,000 Inuit: "The Oceans That We Share: Inuit Nunangat Marine
      Policy Priorities and Recommendations," Inuit Tapiriit Kanatami,
      https://www.itk.ca/wp-content/uploads/2023/03/20230322-Marine
      -Policy-Paper-FINAL-SIGNED.pdf/.
220   stutter into use slowly: Interview with William Halliday, 2021.
220   The biggest source of ship noise: Interview with William Halliday; Ruth
      Teichroeb, "Underwater Noise Pollution Poses a New Threat to Arc-
      tic Wildlife," *Floe Edge Blog, Oceans North,* June 27, 2023, https://www
      .oceansnorth.org/en/blog/2023/06/underwater-noise-pollution-poses-a
      -new-threat-to-arctic-wildlife/; PAME, "Underwater Noise."

220 a mine at Mary River: Joshua M. Jones, "Underwater Soundscape and Radiated Noise from Ships in Eclipse Sound, NE Canadian Arctic," *Oceans North,* January 18, 2021, https://www.oceansnorth.org/wp-content /uploads/2021/02/jjones-eclipse-soundscape-and-ship-noise.pdf/.

220 multiple ships a day: Jones, "Eclipse Sound," 20–27.

220 longest marine coastline in the world: Statistics Canada, "International perspective," *Canada Year Book,* last modified 2016-10-07, https:// www150.statcan.gc.ca/n1/pub/11-402-x/2012000/chap/geo/geo01-eng .htm/.

220 (aviation is another): Interview with Lee Kindberg, 2022.

221 non-binding guidelines: International Maritime Organization, "Addressing Underwater Noise from Ships—Draft Revised Guidelines Agreed," IMO, January 30, 2023, https://www.imo.org/en/MediaCentre /Pages/WhatsNew-1818.aspx/.

221 Marine Strategy Framework Directive: Nathan D. Merchant et al., "A Decade of Underwater Noise Research in Support of the European Marine Strategy Framework Directive," *Ocean & Coastal Management* 228 (2022), https://doi.org/10.1016/j.ocecoaman.2022.106299.

221 Ocean Protections Plan: "Justin Trudeau Announces $1.5B Ocean Protection Plan," CBC News, last updated November 7, 2016, https://www.cbc.ca/news/canada/british-columbia/trudeau-spill -response-1.3840136/.

221 Ocean Noise Strategy: "Mitigating the Impacts of Ocean Noise," Department of Fisheries and Oceans, May 16, 2022, https://www.dfo-mpo .gc.ca/oceans/noise-bruit/index-eng.html/; "Ocean Noise Strategy Roadmap," NOAA, September 2016, https://oceannoise.noaa.gov/sites/default /files/2021-02/ONS_Roadmap_Final_Complete.pdf/.

221 (ANSI): Quantities And Procedures For Description And Measurement Of Underwater Sound From Ships—Part 1: General Requirements, ANSI/ASA S12.64-2009/Part 1 (R2019), (American National Standards Institute, 2019), https://blog.ansi.org/ansi-asa-s12.64-2009 -measuring-ships-underwater-sound#gref.

221 International Organization for Standardization (ISO): Underwater Acoustics, ISO/TC 43/SC 3 (International Organization for Standardization, 2011), https://www.iso.org/committee/653046.html.

221 seismic air guns: Kelly A. Keen et al., "Seismic Airgun Sound Propagation in Arctic Ocean Waveguides," *Deep-Sea Research Part I,* https://doi .org/10.1016/j.dsr.2018.09.0/.

221   a moratorium: "Order Prohibiting Certain Activities in Arctic Off-shore Waters: SOR/2019-280," *Canada Gazette*, July 30, 2019, https://www.gazette.gc.ca/rp-pr/p2/2019/2019-08-21/html/sor-dors280-eng.html/.

221   the west coast of Greenland: J. P. Casey, "The Greenland Freeze: Why Has Greenland Stopped Oil and Gas Exploration?" *Offshore Technology*, August 31, 2021, https://www.offshore-technology.com/features/the-greenland-freeze-why-has-greenland-stopped-oil-and-gas-exploration/?cf-view/.

221   the community of Clyde River: *Clyde River (Hamlet) v. Petroleum Geo-Services Inc.*, 2017 SCC 40, [2017] 1 S.C.R. 1069, https://scc-csc.lexum.com/scc-csc/scc-csc/en/item/16743/index.do/. John Paul Tasker, "Supreme Court Quashes Seismic Testing in Nunavut, but Gives Green Light to Enbridge Pipeline," CBC News, July 26, 2017, https://www.cbc.ca/news/politics/supreme-court-ruling-indigenous-rights-1.4221698/.

222   Gordon Wenz presented a paper: Gordon M. Wenz, "Acoustic Ambient Noise in the Ocean: Spectra and Sources," *Journal of the Acoustical Society of America* 34, no. 12 (December 1962): 1936–1956, https://doi.org/10.1121/1.1909155/.

222   "Combing the white hair": T. S. Eliot, "The Love Song of J. Alfred Prufrock," *Gleeditions*, April 17, 2011, www.gleeditions.com/alfredprufrock/students/pages.asp?lid=303&pg=7. Originally published in *Poetry: A Magazine of Verse*, June 1915, 130–135.

222   over 1,000 kilometers: Wenz, 1962

224   In the Bay of Fundy: Rosalind M. Rolland et al., "Evidence That Ship Noise Increases Stress in Right Whales," *Royal Society* (June 2017), https://doi.org/abs/10.1098/rspb.2011.2429/.

224   prevented Steve Simpson: Interview with Steve Simpson, 2022.

224   Christian Rutz: C. Rutz, et al., "COVID-19 Lockdown Allows Researchers to Quantify the Effects of Human Activity on Wildlife," *Nature Ecology and Evolution* 4 (2020): 1156–1159, https://doi.org/10.1038/s41559-020-1237-z/.

224   (skyrocketing sightings of animals): "The Urban Wild: Animals Take to the Streets Amid Lockdown—In Pictures," *The Guardian*, April 22, 2020, https://www.theguardian.com/world/gallery/2020/apr/22/animals-roaming-streets-coronavirus-lockdown-photos/.

225 Pine happened to be in New Zealand: M. K. Pine et al., "A Gulf in Lockdown: How an Enforced Ban on Recreational Vessels Increased Dolphin and Fish Communication Ranges," *Global Change Biology* 27 (2021): 4839–4848, https://doi.org/10.1111/gcb.15798/.

225 In Glacier Bay National Park: Jan Wesner Childs, "Coronavirus Pandemic Brought Quieter Ocean Waters for Humpback Whales, Other Marine Life," *The Weather Network,* December 10, 2020, https://weather.com/health/coronavirus/news/2020-12-09-quiet-ocean-coronavirus-pandemic-humpback-whales/.

225 Sarasota Bay: E. G. Longden, et al., "Comparison of the Marine Soundscape Before and During the COVID-19 Pandemic in Dolphin Habitat in Sarasota Bay, FL," *Journal of the Acoustical Society of America* 152, no. 6 (December 2022): 3170–3185, https://doi.org/10.1121/10.0015366/.

225 Preliminary anthropause numbers: David R. Barclay, "The Effect of COVID-19 on Underwater Sound," *Journal of the Acoustical Society of America* 150, no. 4 (October 2021), https://doi.org/10.1121/10.0008142. See the dip in container traffic in 2020: "2020 Statistics Overview," Port of Vancouver, accessed September 2023, https://www.portvancouver.com/wp-content/uploads/2021/02/2020-Stats-Overview.pdf/.

226 weeklong traffic jams: Dani Anguiano, "A Record Number of Cargo Ships Are Stuck Outside LA. What's Happening?" *The Guardian,* September 23, 2021, https://www.theguardian.com/us-news/2021/sep/22/cargo-ships-traffic-jam-los-angeles-california/.

226 Bernie Krause was tapped: Interview with Bernie Krause, 2022.

227 "the magnificence of Beethoven's Fifth": Bernie Krause, *Krause's Voices of the Wild* (New Haven & London: Yale University Press, 2013).

227 a 1966 essay: Buckminster Fuller, "The Music of the New Life," *Music Educators Journal* 52, no. 6 (1966): 52.

227 Michael Southworth: Michael Southworth, "The Sonic Environment of Cities," *Environment and Behavior* 1, no. 1 (1969): 49–70, https://doi.org/10.1177/001391656900100104/.

227 credited with popularizing the term: R. Murray Scaffer, *The Soundscape: Our Sonic Environment and the Tuning of the World* (New York: Simon & Schuster, Destiny Books, 1993).

228 The term "niche": G. E. Hutchinson, "Concluding Remarks," *Cold Spring Harbor Symposia on Quantitative Biology* 22 (1957): 415–427. Note that Hutchinson didn't coin the term; it was in use for longer;

see: E. Takola and H. Schielzeth, "Hutchinson's Ecological Niche for Individuals," *Biology and Philosophy* 37, no. 25 (2022), https://doi.org/10.1007/s10539-022-09849-y/.

228   *biophony*: Bernie Krause, "Biophony," *Anthropocene Magazine*, accessed September 2023, https://www.anthropocenemagazine.org/2017/08/biophony/: Interview with Bernie Krause, 2022.

229   (more than 200 meters deep): "What Is the 'Deep Ocean'?" NOAA Ocean Explorer, accessed September 12, 2023, https://oceanexplorer.noaa.gov/facts/deep-ocean.html/.

229   only about 20 percent has been mapped: The Seabed 2030 project has passed the 20 percent mark: Jonathan Amos, "Mapping Quest Edges Past 20% of Global Ocean Floor," BBC, June 21, 2021, https://www.bbc.com/news/science-environment-57530394/.

230   struggling to put a mining code in place: Elizabeth Claire Alberts, "Deep-Sea Mining Rules Delayed Two More Years; Mining Start Remains Unclear," *Mongabay*, July 25, 2023, https://news.mongabay.com/2023/07/deep-sea-mining-rules-delayed-two-more-years-mining-start-remains-unclear/.

229   polymetallic nodules: Sabrina Imbler and Jonathan Corum, "Deep Sea Riches: Mining a Remote Ecosystem," *New York Times*, August 29, 2022, https://www.nytimes.com/interactive/2022/08/29/world/deep-sea-riches-mining-nodules.html/.

230   off the coast of Japan: C. Chen, et al., "Baseline Soundscapes of Deep-Sea Habitats Reveal Heterogeneity Among Ecosystems and Sensitivity to Anthropogenic Impacts," *Limnology and Oceanography* 66 (2021): 3714–3727, https://doi.org/10.1002/lno.11911/.

230   (where ship noise is still quite audible): Robert Dziak, "Ambient Sound at Full Ocean Depth: Eavesdropping on the Challenger Deep," NOAA Ocean Exploration, accessed September 16, 2023, https://oceanexplorer.noaa.gov/explorations/16challenger/welcome.html/.

230   may very well occur in the deep sea: Chen et al., "Baseline Soundscapes."

231   Glass sponges: K. W. Conway et al., "Hexactinellid Sponge Reefs on the Canadian Continental Shelf: A Unique 'Living Fossil,'" *Geoscience Canada* 28, no. 2 (2001): 71–78.

231   recorded the glass reef's soundscape: Stephanie K. Archer et al., "First Description of a Glass Sponge Reef Soundscape Reveals Fish Calls and Elevated Sound Pressure Levels," *Marine Ecology Progress Series* 595 (2018): 245–252, https://doi.org/10.3354/meps12572/.

231 The soundscape framework: Jennifer L. Miksis-Olds, Bruce Martin, and Peter L. Tyack, "Exploring the Ocean Through Soundscapes," *Acoustics Today* 14, no. 1 (Spring 2018): 26–34.

232 species on the glass reef in British Columbia: Archer et al., "First Description."

233 a fire in 1959: Veronica M. Berounsky, "Bay Campus (B)log: The Once and Future Heart of the Bay Campus," University of Rhode Island Graduate School of Oceanography, December 6, 2018, https://web.uri.edu/gso/news/bay-campus-blog-the-once-and-future-heart-of-the-bay-campus/.

233 digitized the original cassette: This seems to be broadly true. For example, the reports by Martin Johnson and others during and following World War II cite "underwater noise" and "biological noise" in their titles. Also, Martin Johnson, "Underwater Noise and the Distribution of Snapping Shrimp with Special Reference to the Asiatic and Southwest and Central Pacific Areas," University of California Division of War Research at the US Navy Radio and Sound Laboratory, UCDWR No U146 Copy no 63, 1944.

233 He digitized the original: Interview with Rodney Rountree, 2021.

234 most common figure: Audrey Looby et al., "A Quantitative Inventory of Global Soniferous Fish Diversity," *Reviews in Fish Biology and Fisheries* 32 (2022), https://doi.org/10.1007/s11160-022-09702-1; Interview with Audrey Looby, 2022.

234 going back to 1874: M. Dufossé, "Recherches sur les bruits et les sons expressifs que font entendre les poisons d'Europe," Annales.des Sciences Naturelles 5, no. 19 (1874): 1–53, and no. 20: 1–134.

234 989 species: Looby et al., "Quantitative Inventory."

235 FishSounds's logo: See www.fishsounds.net.

235 "I'm a bit critical": Interview with Kieran Cox, 2022.

236 "Underwater had always had the qualities": Alex Garland, *The Beach* (New York: Riverhead Books, 1999), 364.

236 Cox got the idea: Interview with Kieran Cox, 2022.

238 Krause is over eighty now: Interview with Bernie Krause, 2022.

238 *The Great Animal Orchestra:* "Exhibition: The Great Animal Orchestra," Fondation Cartier, accessed September 2023, https://www.fondationcartier.com/en/exhibitions/le-grand-orchestre-des-animaux.

238 a million people: Interview with Bernie Krause, 2022.

238 the duo Luftwerk: "Requiem: A White Wanderer," Luftwerk, ac-

cessed September 2023, https://luftwerk.net/projects/requiem-a-white
-wanderer/.

238 Jana Winderen: Carlos M. Duarte et al., "The Soundscape of the Anthropocene Ocean," *Science* 371, no. 6529 (February 2021): 1 of 10,
https://doi.org/10.1126/science.aba4658; Interview with Francis Juanes,
2021.

239 Cascadia Basin: See the basin node lat and long, here: https://www.fdsn
.org/networks/detail/NV/.

239 demonstration neutrino observatory: "Where P-ONE Will Be Located,"
P-ONE, accessed September 2023, https://www.pacific-neutrino.org/p
-one/where-p-one-will-be-located/.

239 stream in from outer space: "What Is a Neutrino?" *Scientific American,*
September 7, 1999, https://www.scientificamerican.com/article/what-is
-a-neutrino/.

239 *Radio Amnion*: Interview with Jol Thomas, 2021.

240 "A Voice Becomes a Mirror": Caitlin Berrigan, "A Voice Becomes a
Mirror Plane Becomes a Holohedral Wand," SoundCloud file, accessed
October 20, 2021, https://soundcloud.com/berrigan/a-voice-becomes-a
-mirror-plane-becomes-a-holohedral-wand/.

240 MSC *Vega*: "MSC Vega," Marine Traffic, accessed October 20, 2021,
https://www.marinetraffic.com/en/ais/details/ships/shipid:755543
/mmsi:636015506/imo:9465265/vessel:MSC_VEGA/.

240 *Maliakos*: "Maliakos," Marine Traffic, accessed October 20, 2021,
https://www.marinetraffic.com/en/ais/details/ships/shipid:755245
/mmsi:636015131/imo:9464247/vessel:MALIAKOS/.

241 Wenz would classify it: Wenz, "Acoustic Ambient Noise."

242 BC Fish Sound projects tested its array: Interview with Xavier Mouy,
2021.

INDEX

ABOUT THE AUTHOR

Amorina Kingdon is a science writer whose work has been anthologized in *Best Canadian Essays* and received honors including a Digital Publishing Award, a Jack Webster Award, and a Best New Magazine Writer from the National Magazine Awards. Previously, she was a staff writer for *Hakai Magazine,* and a science writer for the University of Victoria and the Science Media Center of Canada. She lives in Victoria, British Columbia.